Praise for
THE FARMSTEAD CREAMERY ADVISOR

"Where there's a will, there's a 'whey.' For both the dreamer and those who've already set their sights on the beauty of being a cheesemaker, here is all the information to get started in business and survive. Caldwell's first-hand knowledge takes the reader from their fledgling idea to selling their finished product, and is presented in an easy-to-understand format. Watch out—you may start a venture that just might succeed! This is a brilliant how-to guide, and just what we stewards and entrepreneurs need during a time when our farmland must be saved."
—RICKI CARROLL, owner of cheesemaking.com

"This delightful book is a road map to success for aspiring farmstead cheesemakers. It will help them plan, implement, and develop their new businesses. Passionate stories of experience are revealed, giving great insight into becoming a sustainable, conscientious, and entrepreneurial cheesemaker—including common pitfalls and how best to avoid them. It's about time someone writes such a comprehensive guide! I will recommend this book to every aspiring cheesemaker I know. We would have saved numerous hours and dollars with such an invaluable resource."
—DAVID GREMMELS, President of The American Cheese Society;
co-owner of Rogue Creamery

"*The Farmstead Creamery Advisor* is thorough, eloquent, and generous—a must-have book for anyone considering establishing a creamery. Each point is covered in detail: from the fundamental reasons for going into the business to begin with, to the design of the make room, all the way to the often-neglected exit strategy. This is a good business book for any-sized dairy."
—MAX McCALMAN, author of *Mastering Cheese:
Lessons for Connoisseurship from a Maitre Fromager*

"*The Farmstead Creamery Advisor* is an authoritative, yet friendly and approachable, guide to the process of establishing a farmstead creamery. Simply a must-have for anyone who wants to realize their dream of making and selling cheese."
—TAMI PARR, author of *Artisan Cheese of the Pacific Northwest*

"Here's a nuts-and-bolts, no-nonsense, and essential guide for anyone curious about starting a farmstead dairy. Who better to explain the intricacies and pitfalls of the cheesemaking business than a true practitioner—a woman with a lifetime of experience caring for cows and goats."
—BRAD KESSLER, author of *Goat Song: A Seasonal Life,
A Short History of Herding, and the Art of Making Cheese*

"There have been many books written, and classes given, on the subject of cheesemaking, but primarily from the process-oriented view. Little has been written about how to get started, or the answer to, "What are we up against here?" *The Farmstead Creamery Advisor* fills in those blanks. One of the best pieces of advice in this book is for cheesemakers to build a base foundation beginning with proper business management. No matter how good the cheese, or how much you love your animals, everything depends upon a good business plan to eventually turn a profit. Especially useful, in addition, is Caldwell's chapter on developing an aging space in light of increasing energy costs. This has been a not-so-well-thought-out part of many cheesemaking projects, and could be some of the most important information in this book. To be sure, Gianaclis Caldwell asks the big questions that need to be considered before beginning to develop such a project; it's a true reality check every aspiring cheesemaker needs. In fact, this book should be in all their libraries."

—**JIM WALLACE**, cheesemaking.com

THE FARMSTEAD CREAMERY ADVISOR

THE FARMSTEAD CREAMERY ADVISOR

THE COMPLETE GUIDE TO BUILDING
AND RUNNING A SMALL, FARM-BASED
CHEESE BUSINESS

GIANACLIS CALDWELL

Chelsea Green Publishing
White River Junction, Vermont

Copyright © 2010 by Gianaclis Caldwell.

All rights reserved. No part of this book may be transmitted or reproduced in any form by any means without permission in writing from the publisher.

Project Manager: Emily Foote
Developmental Editor: Benjamin Watson
Copy Editor: Lucy Gardner Carson
Proofreader: Nancy Ringer
Designer: Peter Holm, Sterling Hill Productions

All photographs by Gianaclis Caldwell unless otherwise credited.

Printed in the United States of America
First printing May, 2010

Our Commitment to Green Publishing
Chelsea Green sees publishing as a tool for cultural change and ecological stewardship. We strive to align our book manufacturing practices with our editorial mission and to reduce the impact of our business enterprise in the environment. We print our books and catalogs on chlorine-free recycled paper, using vegetable-based inks whenever possible. This book may cost slightly more because we use recycled paper, and we hope you'll agree that it's worth it. Chelsea Green is a member of the Green Press Initiative (www.greenpressinitiative.org), a nonprofit coalition of publishers, manufacturers, and authors working to protect the world's endangered forests and conserve natural resources. *The Farmstead Creamery Advisor* was printed on Somerset Matte, a 10-percent postconsumer recycled paper supplied by Worldcolor.

Library of Congress Cataloging-in-Publication Data
Caldwell, Gianaclis, 1961-
  The farmstead creamery advisor : the complete guide to building and running a small, farm-based cheese business / Gianaclis Caldwell.
       p. cm.
  Includes bibliographical references and index.
  ISBN 978-1-60358-221-6
  1. Cheesemaking. 2. Cheese industry. I. Title.

SF271.C35 2010
637'.3068--dc22

2010010716

Chelsea Green Publishing Company
Post Office Box 428
White River Junction, VT 05001
(802) 295-6300
www.chelseagreen.com

CONTENTS

Foreword / vii
Preface / xi
Acknowledgments / xv
Introduction / xvii

PART I: *Things to Consider Before Taking the Leap*
    1 What's So Special about Farmstead Cheese? / 1
    2 Making It Official / 16

PART II: *Getting Down to Business*
    3 Sizing Up the Market / 25
    4 Your Business Plan and Company Structure / 37
    5 Production Costs and Issues / 47
    6 Creative Financing / 59

PART III: *Designing the Farmstead Creamery*
    7 Infrastructure and Efficiency / 67
    8 The Milking Parlor and Milkhouse / 82
    9 Cheese Central: The Make Room / 104
    10 Aging Rooms, Cellars, and Caves / 125
    11 Accessory Rooms / 153

PART IV: *Long-Term Survival Guide*
    12 Safety: Why "It Hasn't Killed
          Anyone Yet" Isn't Good Enough! / 161
    13 Increasing Your Bottom Line:
          Classes, Agri-tourism, Additional Products / 172
    14 Keeping the Romance Alive:
          Tips for Re-Energizing / 188

Appendices
    A. Resources / 195
    B. Floor Plans / 203
    C. Milk and Cheese Quality Tests and Parameters / 207
    D. Sample Milk Purchase Agreement / 213

Notes/References / 215
Index / 219
About the Author / 227

Foreword

The renaissance of artisan cheese and other foods continues to develop and spread throughout the United States. Today, growing numbers of craft producers offer us a veritable cornucopia of delicious and sophisticated food products, coupled with their strong commitment to the best environmental practices and fair compensation for farmers and workers. If you enjoy wonderful handmade cheese; world-class beer, wine, hard cider, or spirits; superior-tasting bacon, ham, and sausage; pasture-raised fresh meats and poultry; and organic and sustainably grown fruit and vegetables, then twenty-first-century America is one of the best places in the world to savor these and many more outstanding foods.

During the nineteenth and twentieth centuries, we Americans celebrated our farmers for their amazing skills—tractor and equipment repair, animal husbandry, veterinary medicine, local weather wisdom, and good business sense—as well as for their perseverance, solid democratic values, and plain old-fashioned work ethic. In the years after World War II, however, we transformed agriculture and food production from this human scale to an industrial one in which the farmer's need for diverse skills was supplanted by technology and corporate values.

The damage of this paradigm shift was serious, and it is still being felt today; yet, thankfully, all was not lost. Beginning in the 1970s, American winemakers—some old-timers and a lot of young Turks—took the world by storm and established regions like California's Napa and Sonoma valleys as serious places for winemaking. Following the wine community, craft and home brewers stepped forward and expanded beer culture, with the number of U.S. breweries increasing from a mere forty-four companies in 1980 to 1,302 by 1997. In the beverage sector alone, this proliferation has afforded American consumers choices far beyond anyone's capacity to enjoy them all!

This most recent decade, filled with growth, success, and lessons learned, has assured small business owners and entrepreneurs of continued expansion and opportunity in the agriculture and food sectors. New food businesses appear overnight—a reflection of a national trend toward sustainable agriculture, a desire to know where our food comes from, and a thirst for unique aromas, flavors, and textures. Even in the midst of the current global recession, many consumers demand high-quality foods and are willing to pay a premium for them. More and more chefs and restaurants are passionately devoted to tasty and nutritious—often local—foods, while the local and national media direct their spotlights both on great places to eat and on the people who supply them. Major national food purveyors and small retailers alike are retooling their product lines, renovating buildings to create space for organic items, and teaching staff how to educate patrons about a constantly changing selection of foods.

Beyond the numbers of new food artisans, let's acknowledge their artistry and commitment to excellence, to consistently high-quality beverages and foods. These individuals have resurrected traditional skills, added ideas gleaned from European practices, and mixed them with American ingenuity and innovation to create great products. Today, amplified through the Internet and social media connections, we can access ideas, information, and goods. It is a brave new world in agriculture and food, but not one that's unfamiliar or unwelcome to us. All of these factors—from taste to sustainability to the media—contribute to a powerful public image that attracts the attention of many more people who want to become part of this new revolution. Retirees, second-career individuals, and even some new college graduates are expressing a strong interest in farming or making artisan foods, in connecting with land and animals, and in contributing to healthier communities through taste and flavor.

Artisan cheesemaking follows a similar trajectory. In 1990, approximately 75 small-scale cheese companies existed; a decade later, there were 155 producers. By 2006 I had documented no fewer than 345 artisan cheesemakers, and I estimate that in 2009 the country had more than 425 small cheese companies. Likewise, in the same year the number of craft breweries hovered around 1,500, while dozens of new cured meat producers had joined the ranks of those making traditional bacon and country ham and European-style cured meats.

Since 2000, a number of authors have documented the artisan cheese movement; their books have informed and inspired readers by relating stories about cheesemakers—their histories, their relationships to the land and to animals, and the value of local products to their communities. In my opinion, these artisan cheesemakers have helped to define a sense of place through taste, and they have made important contributions to the future of America's working landscape. They represent a different approach to the country's agriculture, a microcosm of significant contemporary changes that may shape tomorrow's food system.

Through our senses, we experience the links between ecology, economy, people, community, and culture. In a world often characterized by homogenized, standardized foods, artisan cheese represents something distinctive about an area. For cheesemakers, the key is milk: its flavors, color, butterfat, protein, and other elements depend upon myriad factors. Consider for a moment how a whole host of variables—climate and water, geology and soil; geomorphology (the "lay of the land") and pasture; the distinctive breed of dairy animals, what they eat and their care; the season and time of day for milking—adds complexity to the quality and character of milk. The French describe the "taste of place" as *terroir,* all the unique ingredients connected to land and community that distinguish such wines as a Haut Brion from a Latour, a Pomerol from a St. Emilion.

Artisan cheesemaking exemplifies the romance and the reality, the art and the science, of animal husbandry, as well as a devotion to creating distinctive and consistently high-quality products. To create great cheese requires remarkable skills, from scientific and technical matters to business sense and savvy marketing; it also involves an intimate knowledge and understanding of animals, milk, microbiological cultures, *affinage* (cheese aging), hygiene, and sanitation. In short, cheesemaking presents a steep learning curve that even in the best of circumstances might mean not making a profit!

Undoubtedly, some people are attracted to cheesemaking by a vision of bucolic fields, healthy animals, and unique, breakthrough products; yet the reality is generally far from romantic. Today, a potential cheesemaker faces a series of challenges to start up and establish a successful enterprise. Beyond the interplay between artistry and science—creating distinctive cheese styles, while understanding the subtle changes occurring daily in your vat—a prospective cheesemaker must contend with a diverse array of complex needs. From legal requirements and permits to marketing and sales, from facilities design and construction to environmental considerations, a small cheese business must steer through a sometimes exhausting array of tasks.

Gianaclis Caldwell, herself a successful artisan cheesemaker, searched for answers and guides to meet such challenges and demands when she and her husband Vern decided to set up their creamery. She found such books as Ricki Carroll's *Home Cheese Making* and Margaret Morris's *The Cheesemaker's Manual,* which provide some of the practical and technical knowledge needed to create distinctive, consistently high-quality cheese. But no blueprint or map existed to navigate the expensive shoals of permits, the whirlpool of facilities design and construction, or environmental reefs.

The Farmstead Creamery Advisor sets out to fill this critical need. Regardless of size, ambition, or available resources, the stakes are high for a potential cheesemaker. Beyond the recommendation to learn as much as possible about actual cheesemaking from organizations like CalPoly, Wisconsin's Center of Dairy Research, or the Vermont Institute for Artisan Cheese at the University of Vermont, this book steps forward as *the essential* instruction book for new and prospective artisan cheese producers. Gianaclis Caldwell covers all the bases from nascent idea to final product: from decisions about product type to permits and construction; hygiene and sanitation; marketing, distribution, and sales; even labels and logos for your cheeses.

The maturation of the artisan cheese community creates both opportunities and increasing competition. The increasing number of new producers means that there are colleagues with whom to share ideas, ask questions, commiserate, and celebrate successes. On the other hand, with so many artisans at work, often making a variety of cheeses, competition is stiff—and the marketplace becomes less and less forgiving of inconsistency. As food-borne illnesses occur more frequently, wholesalers and retailers demand greater attention to production practices.

Farming, food production, retailing, and restaurants are among the nation's most challenging businesses to launch and sustain successfully. This book is designed to assist a prospective cheesemaker: to be your constant guide, as well as a voice of reason and reassurance that whispers, "The challenges you face are not new; you do not need to re-invent the wheel." By sharing her own story, as well as the experience of other well-known American artisan cheesemakers, Gianaclis invites you to a common kitchen table of knowledge, lessons learned through trial and error, and tested ideas and practices. Beyond artistic expression, in which Pholia Farm stands among the best, this book provides the substance to create a successful, sustainable enterprise . . . and perhaps avoid a few pitfalls along the way.

Congratulations to Gianaclis for making such an important contribution to the future of American artisan cheese and hopefully, in the long run, the nation's agriculture.

JEFFREY P. ROBERTS
author of *Atlas of American Artisan Cheese*
Montpelier, Vermont
January 2010

Preface

First, a confession: Becoming a cheesemaker was never a dream of mine. Growing up, I had no idea that I would one day return to the farm, along with my husband of more than a quarter century and our daughters, to build a small goat dairy. Life often takes you on a journey with an unknown destination. Finding your dream is sometimes as simple as holding on and letting the current lead you to it.

My sister and I were raised with the unspoken—but almost completely enforced—motto of *"If we can't grow it, we can't eat it."* Our family gardened, canned and froze produce, milked cows, and raised cattle for beef and chickens for eggs. We churned butter and made a Greek-style drained yogurt (my mother reports that as a child I would gorge upon this until I was almost sick). My mom tried her hand at aged cheeses too, but there just weren't the information or supplies available back then to help the home cheesemaker learn the ropes, and also there was no one around from the previous generation to pass along the wisdom.

I was a cowgirl: I had my two 4-H Jersey cows, Daffodil and Butterscotch. I sold their milk as a teenager, showed them at the fair, and later became a dairy cattle 4-H leader—but still I had no thoughts of making cheese. I became a licensed practical nurse, not because I wanted to (I wanted to be an artist) but because that was where the jobs were. I tried writing fiction but found that my own talents paled in comparison to the abilities of the authors I loved to read. Eventually, in the late 1980s, I was able to pursue the fine arts of painting and printmaking and take those to what, for me personally, was the apex of expression.

During these years of nursing and printmaking, Vern and I traveled wherever his career in the Marine Corps led us. My cows traveled with me, in a sense—but now they were a herd of porcelain and china mementos given to me by friends and family who understood my love of bovines. We had a horse or two, and often a solitary goat for their companion. I look back on these goats with an apology for not appreciating at the time their uniqueness, both as a species and as individuals. But back then my heart was still with the four-teaters.

At Vern's last duty station with the USMC, we finally owned a piece of property on which we could have a few farm animals. Chickens came first. I have always felt that having a few chickens immediately grounds one to the planet—a circle of life in miniature: you feed them, they feed you. Not to mention, chickens are a hoot!

Author at age fourteen showing her first cow, Daffodil, at the county 4-H fair, 1976.

Finally I started thinking about "something to milk." I wanted our family to have a source of better-quality food. I wanted to not have to use store-bought, factory-farmed milk to make our yogurt. So I went cow shopping. However, the reality of their size, their impact on the land, and the volume of—how shall I say it—poop started to change the way I saw my future with the species.

A neighbor at the time gave me an article from *Mother Earth News* that discussed dairy goats. Now, I had tasted bad goat milk. I had smelled bucks. I didn't think of myself as a goat person. But our youngest daughter wanted to be involved with this milking thing we were embarking on, and the idea of her, at age eight, trying to lead a big cow around was not inspiring. The article mentioned a goat breed that I had not known in my 4-H days, the Nigerian Dwarf dairy goat. They were cute, they were colorful, and best of all, they were little. Amelia was hooked, I was intrigued. The current of our life had shifted.

I started learning how to make cheese before the goats arrived. Using Ricki Carroll's venerable book *Home Cheese Making*, I wowed myself and our friends by making fromage blanc and quick mozzarella. The magic of those first batches is still fresh in my memory.

And then we fell in love with the goats.

A couple of years later Vern was getting close to retiring from the Marine Corps. We knew we could move back home to rural southern Oregon and live on part

Barn at Pholia Farm.

of my family's original farmland. But what kind of work could Vern find that would both be fulfilling for him and allow me to stay at home, make cheese, ride the horses, and "do art"? By this time I was making hard cheeses, teaching some beginning classes, and having a ball. Vern had always been a cheese nut and a lover of all dairy products, so he had no complaints. Amelia and our older daughter, Phoebe, didn't seem to mind helping devour the spoils of my labor, either. Amelia and I had an idea: We could move back to Oregon and start a little cheese dairy. Vern was all for it.

We were swept up by the intensity of the venture and have been holding on ever since—sometimes thrilled, sometimes wanting to jump ship, but for the most part amazed and fulfilled.

In the process of writing this book I have satisfied some of my own cravings, for knowledge and for expressing myself through writing. Some questions that we still had about the business after all these years have been answered, and other new ones have been raised. I hope the wisdom and knowledge shared with me for this book by cheesemakers from across the country will help those whose journey to cheesemaking is just beginning, and even those of us who are still on board and enjoying the ride.

Bon voyage!

*"I may not have gone where I intended to go,
but I think I have ended up where I intended to be."*
—Douglas Adams, 1952–2001

Acknowledgments

Thank you to my dear, longtime friend Christine DeMont; to good friend, author, and cheese advocate par excellence Tami Parr; and to my super-supportive husband, Vern, for the much-needed encouragement, editing, support, and advice. To my youngest daughter, Amelia: Thank you for pitching in with extra milking and goat chores so I could spend evenings writing and take several jaunts across the country to visit other farmstead creameries. When you are working on your book, I will do the same!

Thank you to the cheesemakers who allowed me to pester them with questions and/or visits: Alyce Birchenough and Doug Wolbert, **Sweet Home Farm**, Alabama; Kelli and Anthony Estrella, **Estrella Family Creamery**, Washington; Jan and Larry Neilson, **Fraga Farm**, Oregon; Marjorie Susman and Marion Pollack, **Orb Weaver Farm**, Vermont; Pat Morford, **River's Edge Chèvre**, Oregon; Brad and Meg Gregory, **Black Sheep Creamery**, Washington; Evin Evans, **Split Creek Dairy**, South Carolina; Leslie Cooperband, **Prairie Fruits Farm**, Illinois; Robin Clouser, **Mama Terra Micro Creamery**, Oregon; Walter and Liz Nicolau, **Nicolau Farms**, California; Laurie and Terry Carlson, **Fairview Farm**, Oregon; Jennifer Lynn Bice, **Redwood Hill Farm**, California; Rhonda Gothberg, **Gothberg Farms**, Washington; Gayle Tanner, **Bonnie Blue Farm**, Tennessee; Michael Lee, **Twig Farm**, Vermont; Laini Fondiller and Barry Shaw, **Lazy Lady Farm**, Vermont; Willow Smart and David Phinney, **Willow Hill Farm**, Vermont; Brian Futhey, **Stone Meadow Farm**, Pennsylvania; Linda and Larry Faillace, **Three Shepherds**, Vermont; Willi Lehner, **Bleu Mont Dairy**, Wisconsin; Paul and Kathy Obringer, **Ancient Heritage Dairy**, Oregon; David and Kathryn Heininger, **Black Mesa Ranch**, Arizona; Angela Miller, **Consider Bardwell Farm**, Vermont and to all of the other cheesemakers whose lives and cheese have inspired and influenced my own cheesemaking and writing.

I am very grateful to the following experts who reviewed pertinent chapters for technical accuracy: Dr. Lisbeth Goddick, **Oregon State University**; Paul Hamby, **Hamby Dairy Supply**; and Peter Dixon, **Dairy Foods Consulting**.

And last, thank you to all those who have worked for years to help small cheesemakers find their way, including Vicki Dunaway, author, editor, and publisher of the quarterly publication *CreamLine* and the **Small Dairy** website (www.smalldairy.com); Peter Dixon, publisher of the *Farmstead Cheesemaking* periodical, consultant, and teacher; and Ricki Carroll, **New England Cheesemaking Supply**, pioneering cheesemaking author and supplier/resource.

And finally, one more thank you, to everyone at Chelsea Green Publishing. While this was my first experience working with a publisher, I cannot imagine it getting any better. Their efforts, enthusiasm, and engagement made it obvious that each book they produce is as important to them as it is to its author.

Introduction

There are many paths to the career of farmstead cheesemaker. Some of us simply seek a return to the farm and the "simple" life; others have found fulfillment in making pure, artistic cheeses; still others are seeking ways to make their animal hobby pay for itself. However you have arrived at this point—where the business of farmstead cheese sounds appealing—this book was written with you in mind.

When we at Pholia Farm were going through the process of designing our small creamery, it was a near impossible challenge to find the resources that would guide us in the right direction. In fact, like so many other cheesemakers, many of our correct choices were made only after making incorrect, often costly decisions. While the logical place to start looking for answers is with other successful cheesemakers, the reality of their busy lives often leaves them too busy to support the frequent requests for mentoring. And that's where this book fits in. *The Farmstead Creamery Advisor* leans heavily upon the shared wisdom of these experienced farmer-cheesemakers to provide a valuable knowledge base for both aspiring as well as established cheesemakers.

Let me tell you first what this book will *not* do: It will not teach you how to make cheese or how to manage dairy animals—that information is thoroughly covered by other books and resources (many of which are listed in the back of this book). I see the business of cheesemaking as a triangle, with one side representing cheesemaking; the other side, animal husbandry; and the base of the triangle, business. Without a strong and stable base, these other skills will collapse under the pressure of the free enterprise universe. *The Farmstead Creamery Advisor* will help guide you through all the aspects that are involved in getting started as a commercial farmstead producer—from designs and floor plans, permitting and regulations, and equipment and setup to marketing and sales, and even saving money through efficient use of energy and time.

Part 1, "Things to Consider Before Taking the Leap," begins with a brief history of artisan cheesemaking in the United States, to give you some appreciation for the amazing growth in the industry in just the past three decades. The second chapter provides an overview of some of the important issues you will want to address early on in your process, including zoning, permits, and a slightly tongue-in-cheek "suitability quiz."

In part 2, "Getting Down to Business," we'll take an in-depth look at everything the farmstead cheesemaker should know about such topics as marketing, business plans and structure, production costs and issues (including insurance, labor, and product loss), and how to finance your operation.

Part 3, "Designing the Farmstead Creamery," is the heart of the book. It begins with an overview of infrastructure and efficiency choices that you should be aware of throughout the entire design process. Then it covers in detail the main rooms of the farmstead creamery—milking parlor, milkhouse, and make room—from the standpoints of design, construction, regulatory issues, and equipment options. Finally, aging rooms, cellars, and caves, as well as other accessory rooms you might want in your creamery, are discussed. This entire section is sprinkled with pertinent tips and examples from other cheesemakers.

Part 4, "Long-Term Survival Guide," begins with an approachable look at food safety and how the small cheesemaker can implement a workable quality control program. Then we'll talk about some of the myriad options that exist for increasing your financial bottom line, including agri-tourism, classes, and such products as fluid milk and meat. The section ends with a chapter whose subject is near and dear to me: sustaining your passion for the job—a topic that I feel is often underemphasized, but one that is incredibly important if you are to be successful over the long haul.

Finally, the appendices contain some useful additional material for the home and commercial farmer-cheesemaker: a list of resources; several floor plans (including a small, home dairy plan); milk and cheese quality information; and a sample milk purchase agreement.

While the number of people building and starting farmstead creameries continues to rise, the number of people leaving the field—often after an all-too-brief career—is also increasing. My goal in writing this book is to provide a guide filled with practical information and tools to help you create and sustain the best possible business and long-term future as a farmstead cheesemaker.

PART I
THINGS TO CONSIDER BEFORE TAKING THE LEAP

· 1 ·

What's So Special about Farmstead Cheese?

The United States is experiencing a food-quality renaissance. An increase in the number of farmers' markets and "eat local" campaigns, a growing awareness of food quality, and a desire to appreciate the story behind the product are all influencing the way Americans are buying and consuming food. While we are still largely a nation of fast-food addicts and all-you-can-eat buffet aficionados, more and more people today are starting to care less about the size of the serving than about the quality and story of its ingredients. This awakening is not limited to those who can afford the luxury of finer foods. It extends—and indeed originates—from a basic need to reconnect with health, history, and the awareness of nutrition's role in our very existence.

The History of Cheesemaking in the United States

Bernard Nantet, in his book *Cheeses of the World,* maintains that the United States, unlike Europe, does not have a strong tradition of artisan cheesemaking. It could be argued that it is this lack of an embedded culinary-cultural background, in part, that allowed the unfettered mechanization that all but extinguished the manufacture of handcrafted artisan cheeses in the U.S. by the mid-1900s. The current revival, which began in earnest in the late 1970s, occurred thanks to a combination of factors that increased the American public's appreciation not only of food but also of the way of life that the farmer-cheesemaker leads.

Rise and Fall

Although goats, sheep, and cows traveled to the Antilles (Caribbean islands) with Christopher Columbus in the late 1400s, it wasn't until the early 1600s that milk cows, and along with them cheesemaking, arrived at European settlements on the shores of what is now the United States of America. Cheeses were part of the provisions stocked on board ships traveling to the Americas, and as with all foods packed for the difficult voyages, cheese was a sustenance food, not a luxury.

Cheese, both on board the ships and in the new settlements, was simply the best way to preserve excess milk and extend the availability of a valuable food.

European immigrants adapted to the hardships of life in the New World while continuing to practice the food traditions of their native cultures. Over time and through continued waves of immigration, cheese produced in America gradually began to reflect regional influences: In the northeast part of the country, an English influence created an early Cheddar industry; in Wisconsin, Swiss and Danish traditions included Gouda and alpine styles; and in California and the West, Spanish and French cultures influenced the kinds of cheeses made there, including the development of an American original, Monterey Jack cheese. By the mid-1800s most rural families had a milk cow or goats for dairy, meat, and by-products. Cheese was produced on the farm or at home, and cheesemaking was a normal part of a homemaker's repertoire. The seeds of change, for all of agriculture and eating, came with the American Industrial Revolution in the 1850s. Mechanization increased the ability of farmers to grow more feed, raise more animals, and subsequently harvest ever-increasing quantities of milk. For the cheesemaker, equipment could be manufactured to process larger volumes of milk into cheese to feed a growing population.

In the 1840s a Wisconsin man named James Picket is believed to have been the first farmer to make cheese from the milk of not only his own animals, but a neighbor's cows as well. This new concept in dairying was taken a step further in 1851 when the first "modern" cheese factory was built by Jesse Williams in Oneida, New York. Williams's factory is believed to have been the first cheese plant to pool milk from multiple farmers and complete the entire process of cheesemaking in a commercial facility. Other factories quickly sprang up

Family cow being milked, early twentieth century.

throughout the country. By 1880 there were 3,923 factories nationwide, with a production volume of 216 million pounds of cheese. The family cow was on her way out of the picture.

By the 1920s cheese production had reached 418 million pounds, with most of this still occurring in what would be, by today's standards, small to moderate-size facilities processing milk from only local dairies as well as their own milk. By the 1930s cow's-milk cheeses similar in style to most major European cheeses were being made at the industrial level.

The early part of the 1900s also saw the birth and infancy of what would become the modern-day super-mega-one-stop grocery store. Previously, shopping had been done at specialized stores—the butcher, the baker, the green grocer. But by 1910 many stores began carrying multiple specialty foods under one roof. This consolidation of products led to the building of ever-larger stores, the development of chain stores, and the need for centralized distribution. The competitive drive to promote the cheapness and value of one supermarket over another quickly followed. These factors all contributed to the impetus to produce cheese in greater volume and in the most cost-effective manner possible. Americans began to compromise quality for pocketbook "value."

The Great Depression of the 1930s brought further woes to the small producer. While many small dairies folded under the economic strain, others survived, in part thanks to the formation of cooperatives, as well as the intervention of creameries that refocused their production to purchase their fluid milk from these struggling small farms (see sidebar).

> **THE VELLA FAMILY**
>
> Tom Vella founded Vella Cheese in 1931 in Sonoma, California, with a commitment to local dairymen. His creamery helped many small dairy farms stay in business during both the Depression and World War II. Vella's successful efforts extended beyond Sonoma to southern Oregon, where he started the Rogue Creamery in the 1930s. Both the Rogue Creamery (now owned by Cary Bryant and David Gremmels) and Vella Cheese (owned by Tom Vella's son, Ignazio) continue today with a focus on local sustainability as well as on the creation of artisan cheeses that are rooted in tradition.

Following on the unfortunate heels of the Depression, World War II furthered labor and economic issues by its upheaval of the workforce (men left farms and factories for the battlefield) and the necessary redistribution of resources and supplies to the war effort. When the conflict finally ended, wartime technological advances transitioned to civilian-oriented purposes. The increased technology available to manufacturing, combined with the demand for cheaper and more modern products (often seen as superior by a population starved for finer goods at an affordable price), spelled trouble for the small handmade-cheese producer.

Revival

The re-emergence of the small cheesemaker began in earnest in the 1980s. As with the decline of handmade cheese, the renaissance occurred in response to the influence of movements and trends that occurred in the twentieth century.

> **GOURMET OR FOODIE?**
>
> A gourmet is an aficionado of fine food and dining. The term "foodie" was first used in the 1980s to refer to a person whose hobby and passion center around everything to do with food, including understanding its manufacture and preparation. This new avocation in our culture has contributed not only to increased sales of small-production cheese, but to diverse income possibilities for farmer-cheesemakers through agri-tourism, cheesemaking classes, etc. For more on value-added agri-tourism, see chapter 13.

Hippies, back-to-the-landers, and gourmets (see sidebar) prepared the way for the renaissance of handmade cheese.

Occurring almost simultaneously, and running different but overlapping courses, the hippie and the back-to-the-land movements both peaked in the 1960s through the mid-1970s. Their roots are vastly different, but their influence on the awareness of food quality and its effects on health and happiness are similar. The hippie movement brought an interest in natural and "health" foods, while the back-to-the-landers sought a return to the agrarian and self-sufficient lifestyle of their forebears.

The back-to-the-land movement saw the return of many urban and suburban dwellers to the countryside. The concept of homesteading brought renewed interest in the family milk cow and dairy goat. Beginning in the 1970s—and still going strong today—the magazine *Mother Earth News* and the Foxfire book series provided guidelines and inspiration for rural living and self-reliance. For many people, the homesteading spirit and lifestyle proved to be a transient state, once the hardships and reality of "living off the land" hit home. But even those who went back to more modern lifestyles did not lose the appreciation for that way of life.

While some parts of our society were interested in reconnecting to the land, a more traditional way of life, and the quality of food that lifestyle offered, another segment was developing a culinary consciousness that included an expanding appreciation of food flavors and quality. Increased and easier travel to Europe, especially France, exposed many to flavors and cooking that had been ignored, for the most part, in the modern American diet. This appreciation was helped immensely by the work of such people as Julia Child, whose book *Mastering the Art of French Cooking* and television show *The French Chef* helped many mainstream Americans develop a new interest in the quality of their food, and Alice Waters, chef and proprietor of the Berkeley, California, restaurant Chez Panisse and a leader in the Slow Food movement (see sidebar).

Farmstead cheese pioneer Jennifer Lynn Bice, Redwood Hill Farm, California, making cheese at her on-farm plant in the late 1980s.

SLOW FOOD AND AMERICAN RAW-MILK CHEESE

Slow Food is a non-profit, member-supported international organization that was founded in Italy in 1989 to counteract fast food and fast life, the disappearance of local food traditions, and people's dwindling interest in the food they eat, where it comes from, how it tastes, and how our food choices affect the rest of the world. Slow Food works to defend biodiversity in our food supply, promote "eco-gastronomy" and taste education, and connect producers of excellent foods with consumers (or "coproducers," as the organization describes them) through various events and initiatives.

One of Slow Food's main international events is the biennial "Cheese" gathering that takes place in Bra, a market town in the Piedmont region of northwest Italy, just south of Turin. At Cheese, according to Slow Food's website, "the world's most renowned artisans, affineurs [cheese agers], cheesemongers, and shepherds come to present their cheeses to tens of thousands of visitors and host taste workshops."

Slow Food has a strong presence in the United States through its national organization, Slow Food USA (www.slowfoodusa.org). The U.S. organization has several programs that feature artisan cheese, including the American Raw Milk Cheese Presidium, which is designed to recognize unique and valuable U.S. cheeses.

The Raw Milk Cheesemakers' Association (RMCA) was founded under the guidance of Slow Food USA but operates independently. RMCA promotes the production of high-quality raw-milk cheeses by providing criteria and guidelines for its membership that support both a quality product and humane production methods. See the resources and notes for contact information for the RMCA.

As all of these influences converged, a market for artisan, American-made cheese began to develop and a new wave of pioneers rose to meet the call. Cheesemakers, authors, educators, and visionaries have all had a hand in the current success of handmade cheese in the United States. Here are just a few of these pioneering farmstead cheesemaker innovators and leaders: Laura Chenel, Laura Chenel's Chèvre, California, 1979; Sally Jackson, Sally Jackson Cheese, Washington, 1979; Allison Hooper and Bob Reese, Vermont Butter and Cheese, Vermont, 1984; Judy Schad, Capriole, Indiana, 1988; and Jennifer Bice, Redwood Hill Farm, California, 1988. (Of these, Capriole and Sally Jackson remain farmstead operations.) Authors such as Laura Werlin (who has been writing about cheese in articles and books since 1999) and Max McCalman (whose books and speaking engagements have helped elevate the role of cheese in fine dining and the status of cheesemongers and maître fromagers) have greatly increased the public's awareness and appreciation of cheese, as well as its makers. Educators and visionaries include Ricki Carroll, author and cofounder of New England Cheesemaking Supply in 1978, who continues to provide supplies and education to cheesemakers—home, hobby, and professionals alike; Frank V. Kosikowski, founder of the American Cheese Society (see sidebar on next page) in 1983 and author of *Cheese and Fermented Milk Foods;* and Paul Kindstedt, coauthor with the Vermont Cheese Council of *American Farmstead Cheese* and an original member of ACS. It is thanks to these leaders, as well as many others, that the way has been paved for the many new cheesemakers who are experiencing such success today.

> ## THE AMERICAN CHEESE SOCIETY
>
> According to its website, "the American Cheese Society was founded in 1983 by Dr. Frank Kosikowski of Cornell University as a national grassroots organization for cheese appreciation and for home and farm cheesemaking."
>
> **Mission and Purpose**
> 1. To uphold the highest standards of quality in the making of cheese and related fermented milk products.
> 2. To uphold the traditions and preserve the history of American cheesemaking.
> 3. To be an educational resource for American cheesemakers and the public.
> 4. To encourage consumption through better education on the sensory pleasures of cheese and its healthful and nutritional values."
>
> As of 2009 the American Cheese Society had 1,235 members, of whom 409 are cheesemakers (ACS does not differentiate between farmstead and other cheesemaker memberships). For the 2009 ACS Judging and Competition, 1,327 cheese and dairy products were entered.

Defining the Small, Farmstead Cheesemaker

Now that you know some history of farmstead cheesemaking in the United States, let's talk about some definitions, motivations, and qualifications.

"Artisan," "Farmstead," and Production Size

The term "artisan" is applied to any product (food or otherwise) that is made in limited quantities by a skilled craftsman, usually by hand. The term is not legally defined for business use, however, and is becoming another buzzword whose meaning is being diluted by overuse. The American Cheese Society does define "artisan" when applied to cheese (see sidebar). "Artisan" and "artisanal" (interchangeable terms) imply, but do not guarantee, high-quality products!

"Farmstead" is a term applied to cheese made only from the milk of the farmer's own animals; the term "farmhouse" is sometimes used interchangeably, but it is not as common. The production size of a farmstead cheese business is not limited or defined. In consumers' minds, however, it is often assumed that the facility is small and not highly mechanized. The farmstead cheesemaker is usually the smallest size of cheese producer, but not always. One very successful farmstead creamery in Wisconsin milks (according to its website) a herd of

> ## ARTISAN OR ARTISANAL CHEESE
>
> The word "artisan" or "artisanal" implies that a cheese is produced primarily by hand, in small batches, with particular attention paid to the tradition of the cheesemaker's art, and thus using as little mechanization as possible in the production of the cheese. Artisan, or artisanal, cheeses may be made from all types of milk and may include various flavorings.
>
> In order for a cheese to be classified as "farmstead," the cheese must be made with milk from the farmer's own herd, or flock, on the farm where the animals are raised. Milk used in the production of farmstead cheeses may not be obtained from any outside source. Farmstead cheeses may be made from all types of milk and may include various flavorings.
>
> Source: The American Cheese Society (www.cheesesociety.org)

TABLE 1-1: Creamery Size Ranges		
Very Small	**Small**	**Medium**
Under 10,000 lbs cheese/year	10,000–20,000 lbs cheese/year	20,000–100,000 lbs cheese/year
20 ewes* = 2,040 lbs cheese	100 ewes* = 10,200 lbs cheese	200 ewes* = 20,400 lbs cheese
15 does** = 3,750 lbs cheese	50 does** = 12,500 lbs cheese	100 does** = 25,000 lbs cheese
5 cows*** = 5,000 lbs cheese	10 cows*** = 10,000 lbs cheese	20 cows*** = 20,000 lbs cheese
* East Friesian ewes producing 600 pounds of milk each per year. Assumes a 17% cheese yield. ** Alpine does producing 2,500 pounds of milk each per year. Assumes a 10% cheese yield. *** Guernsey cows producing 10,000 pounds of milk each per year. Assumes a 10% cheese yield.		

950 Holstein cows, whose production level allows it to make approximately 3 million pounds of cheese annually. Many other existing cow dairies have value-added cheese plants in which they produce their own farmstead cheese. Cheese is saving many a family farm in this fashion.

Another term you will see is "specialty" cheese. Specialty cheese is produced by large-scale, industrial cheese companies as a value-added product of higher quality and in a limited quantity as compared to their other cheese products. According to the Wisconsin Specialty Cheese Institute, a specialty cheese cannot exceed an annual nationwide volume of more than 40 million—yes, *million!*—pounds. Both artisan and farmstead cheeses sometimes fall under the category of specialty cheese when being discussed in industry trade papers.

This book focuses on the small and very small cheese business. Table 1-1 defines the size of a creamery based on its annual production of cheese. I am providing these definitions to help give prospective cheesemakers some idea of the size and scope they will be looking at in order to meet their production goals. At this time, the American Cheese Society has not formally defined these terms. Also, keep in mind that the production data are based on estimates and averages only. Actual yields will vary greatly based on breed, management, type of cheese, and individual animal differences. Remember, these numbers are just to give you an idea of what size dairy you might want to consider. When you are considering the size of your business, both the number of animals as well as the production volume of cheese must be considered when formulating your business plan. We will talk more about this in part 2, "Getting Down to Business."

The Motivation behind Becoming a Farmstead Cheesemaker

Farmstead cheesemakers are usually a unique blend of farmer, animal lover, independent spirit, masochistic laborer, and artist. Very few choose this life with monetary goals as their number-one motivator; instead it is a passion for the animals and for a way of life, the desire to create a value-added product on an existing farm, or the desire to leave a prior profession or lifestyle for the pursuit of a more rural way of living. There are also people who enter the business with purely entrepreneurial motivations—those for whom the growing prestige and market potential of artisan cheese is the magnet (not unlike the motivation that draws some to plant a vineyard or build a winery). But, for the most part, the farmstead cheesemaker is first and foremost a herdsman. Let's take a look at some of the most common reasons for building a farmstead creamery, along with the assets and pitfalls that each motivation brings to the mix.

Hobby to Profession

Those who start out with a couple of dairy goats or a milk cow to feed a growing family, or a desire to live a more self-sustaining lifestyle, often find themselves with more milk than they know what to do with. Learning to make cheese is a

Pholia Farm today.

logical progression that becomes a gratifying hobby for many. The decision to "go pro" is sometimes seen as an elaboration of the hobby, when in fact it is truly a full-fledged transformation. The change from avocation to profession brings new dimensions that can wear out even the most passionate hobbyist.

- *Assets:* Existing animal management skills, awareness of the rigors of farm life, some cheesemaking skills.
- *Pitfalls:* Often a lack of business management training, possible lack of investment capital.

Value-Added

For many dairy people, adding a cheese facility to the dairy farm provides a value-added product that increases the prospects for survival of the farm. The growing popularity and public perception of cheese is helping retain and bring back generations of family that might not have previously stayed on the farm.

- *Assets:* Existing animal management skills, awareness of the rigors of farm life, existing business structure, existing infrastructure (buildings, systems, etc.).
- *Pitfalls:* Possible lack of cheesemaking skills and a lack of time to leave the farm for training.

Career and Lifestyle Change

Whether a long-contemplated dream or a recent revelation, more and more people are launching farmstead creameries after leaving their previous careers. Often these careers had little to do with the day-to-day operations of a farm, but maybe they brought them into contact with fine cheese or a rural, agrarian way of living. As often as not, the career change is a return to roots or a family history after experiencing the "regular" work world.

- *Assets:* Possible business skills, investment capital, and/or a retirement income.
- *Pitfalls:* Possible lack of animal experience, possible lack of physical stamina related to age at retirement.

The Entrepreneur

For investors building an artisan cheese business, the need for a reliable source of the highest-quality milk often leads them to the farmstead solution. The size and scale of these operations is medium to large. Usually both herd managers and cheesemakers are employed to handle these parts of the operation.

- *Assets:* Financial resources, business skills.

- *Pitfalls:* Lack of animal management expertise. (Cheesemaking experience, I believe, is more readily learned than animal husbandry—educational opportunities and professional expertise are easier to find than an in-depth animal husbandry education.)

The Hybrid

Many farmstead cheesemakers are a mixture of some or all of the above motivations. People entering the industry with a varied background and multiple inspirations often bring a mix of qualifications that promote success in ways that cannot be anticipated by simply analyzing their credentials. There is no way to accurately categorize this type of person, but it is still important for them to attempt to analyze their skill set and job suitability based on information gained while researching the industry.

Do You Really Want to Do This?

It seems like being a farmstead cheesemaker would be fun and fulfilling, but once you take a good, hard look at the realities of setting up and running your own creamery, you need to decide if it is the right move for you. Here is a little quiz, devised with the help of cheesemakers from across the country, to help set the stage for what you will be in for should you bravely go where others have gone before (despite their warnings!).

Let's look at these questions in more detail. If it seems a bit discouraging, try to remember that many of these issues will not seem as daunting after you learn more. The knowledge and skills you will gain by reading this book and educating yourself through other opportunities will give you the tools you need to deal with each of these issues, should you choose to become a farmstead cheesemaker.

Are the hours really that bad? There are times throughout the year when most farmstead cheesemakers find themselves going to bed just about in time to get up again. Kidding/lambing/calving season is a prime example—and this is also the same time that most farmers' markets start their season. Milk is flowing, cheese must be made, and babies won't wait for your bedtime schedule. It is often nonstop work, and you feel like you're never caught up. When you choose to become a

---

TEN QUESTIONS TO TEST YOUR SUITABILITY

1. Do you like to get up early—every day of the year and for many years to come?
2. Do you mind working late into the evening—and then getting up early the next day?
3. Do you mind working hard between getting up early and going to bed late?
4. Does your spouse or partner also enjoy these hours?
5. Do you have a good head for business?
6. Do you have an artistic or creative flair?
7. Can you be satisfied with repetitive labor and a lot of dishwashing?
8. Do you have a great love for working with animals, no matter how exhausted you are?
9. Can you deal well with constantly changing challenges and problems, including animal deaths, equipment failure, product loss, possible lawsuits and product recalls, and rising insurance and power costs?
10. Do you mind working for below minimum wage for several years, or do you have an independent source of income to help pay bills?

An artistically designed exterior helps make Pholia Farm's Hillis Peak an eye-catching cheese.

farmstead cheesemaker, you are choosing not just a job, but a way of life. If you have a spouse or partner, you will need to consider very carefully whether or not this way of life will be fulfilling for both of you, together.

How about a good head for business? When the hobby farmer-cheesemaker turns pro, everything changes. In reality you are now operating two businesses—a dairy farm and a cheese business. Any inefficiency in either aspect will likely evolve into a liability, both financially and, in the end, emotionally. If you know you will not be able to develop a sound business plan, maintain accurate and up-to-date financial books, complete invoices, and follow up on orders and billing—and you still want to go into the business—then consider taking classes, or even hiring a bookkeeper and office manager.

Why would I need to be creative or artistic? Remember there is "art" in "artisan." Not only will being creative give you an edge in producing visually appealing products, but it will help with designing packaging, labels, and promotional materials. As the number of producers grows and the volume of farmstead cheeses increases, it will be the little things, such as irresistible packaging and mouthwa-

> **TEN CHARACTERISTICS THAT MAKE FOR A GREAT FARMSTEAD CHEESEMAKER**
>
> 1. Organizational skills
> 2. High energy level
> 3. Personal drive
> 4. Ability to delay gratification
> 5. Patience
> 6. Kindness
> 7. Persistence
> 8. Creativity
> 9. Work ethic
> 10. Ingenuity

tering product presentation, that will help give your business an edge.

Is there really a lot of dishwashing and repetitive labor? Oh my, yes! Once all of your cheese recipes have been refined and perfected, it becomes the great cheesemaker's job to keep making them, as identically as possible, over and over. Keeping the passion and inspiration evident in each batch and wheel can become a challenge. As to dishwashing, there is a standard saying that cheesemaking is 90 percent cleanup. Sanitation and cleanliness in a licensed creamery cannot be treated casually. It is not in the least bit glamorous or inspiring, but you will spend a good deal of time doing it.

How could I be too tired to enjoy my animals? For most farmstead cheesemakers the animals are usually the reason they make cheese, not the other way around. Once you are licensed, however, selling your cheese becomes a priority that can take away time with the animals and drain your patience and energy to deal with their needs, as well as the challenges that caring for them brings. It's not hard for the pressures of the cheesemaking side to leach the joy out of the original reason for starting the business—the animals.

What kinds of problems can crop up? The farmstead creamery, no matter how well administered, will face an ever-changing set of challenges. Dealing with equipment failure that leads to lost production or lost product; animal health

> **PHOLIA FARM: OUR MOTIVATIONS**
>
> We are definitely in the hybrid category! While I had a rural upbringing and a background in dairy cattle, Vern had a more "rural-suburban" background in addition to his professional military career (including the Naval Academy and a master's degree in contracts and acquisitions). I was a nurse (an LPN and LVN for the better part of a decade) and then a professional artist. Vern's retirement from the military, available land from my family, and the desire to create a sustainable lifestyle led us to dairy goats, which in turn led to cheesemaking and ultimately to the "retirement" career choice of a farmstead creamery.
>
> Our diverse backgrounds brought assets that we could not have anticipated, such as business skills, decision-making capabilities, and an artistic approach to our products. Our past choices over the years regarding investments yielded enough capital to make the initial investment in our creamery. Vern's retirement income, while not a lot, also provides a safety net that makes sure the essential bills can be paid during the years that our cheese business is still in a fledgling state.
>
> I see our motivation as a definite hybrid: my hobby as a kid with dairy cattle has turned into a profession; we took a chunk of family land that was dormant and added value to it. This was most certainly a huge career and lifestyle change, and we feel that we have entered a market that appeals to our entrepreneurial natures.

Cheesemaking class at Oregon State University. Photo by Lynn Ketchum, courtesy of Oregon State University Extension and Experiment Station Communications.

issues that lead to lost milk, animal deaths, and culling decisions; and the possibility of liability lawsuits, product recall, and inspection violations—all these and more bring a facet to the lifestyle that can be unduly stressful. To be successful, you must be prepared to face these challenges without letting them overwhelm you.

What about money? Even if your cheese sells at the high end of the price spectrum, the number of hours you will work to create that product could mean that your average income will be somewhere below minimum wage—*I am not kidding.* If you do not have the investment capital to survive the first few years, or another source of income to make ends meet, then you would be wise to reconsider starting a cheesemaking business (or any small business, for that matter). Even after several good years, you will probably not become wealthy making cheese—but you will have a priceless quality of life and hopefully be able to pay the bills!

Some of these questions may seem extreme, but the reality of the lifestyle of a farmstead cheesemaker is at times very difficult and intense. If you answered yes, even if it was a somewhat reluctant affirmative, to *all* of the questions in the quiz, then you are quite likely well suited to the profession of farmstead cheesemaker. But if you have any hesitation in embracing these conditions as a huge part of your life, then I would encourage you to enjoy this book, tour cheese farms, eat farmstead cheese, make your own cheese at home—as a hobby—and have a life!

Learning the Craft

So where do you learn how to make cheese? Most start learning when inundated with pails and pails of milk—in other words, out of necessity. But when the hobby is about to become a profession, other resources should be explored. Learning the art of cheesemaking, as well as the science and safety behind the process, through experienced teachers will help ensure your success as a business.

There are several venues in which to learn both the art and the science of cheesemaking (see appendix A for a list of resources for cheesemaker educational opportunities):

- Books, Internet
- University short courses
- Private workshops and classes given by cheesemakers and educators
- Apprenticeship/internship programs at working farmstead creameries
- Traveling to other countries with strong cheesemaking traditions to research traditional practices

In most states a business can be a licensed cheesemaking facility without the proprietor having special training as a cheesemaker, but in some there are standardized requirements. For example, in order to obtain a cheesemaker license in the state of Wisconsin, special training regulations apply, including up to 18 months of on-the-job work as a cheesemaker assistant. Be sure to investigate your state's laws in this regard. (More on this in chapter 2, "Making It Official.")

Many cheesemakers continually seek to expand their knowledge and mastery of the craft long after obtaining a license. Entering competitions, seeking technical reviews of their cheese, taking courses, subscribing to professional publications,

LIFELONG LEARNING

Alyce Birchenough, founder and cheesemaker of Sweet Home Farm in Elberta, Alabama, has been making her award-winning raw-milk cheeses for more than twenty years. When Alyce and her husband, Doug Wolbert, built their small cow dairy in the mid-1980s, there were not as many options available for learning as there are now. Alyce found her first information through the Minnesota Farmstead Cheese Project, an extension service project that was designed at the time as a value-added opportunity for farmers. She also relied upon back-to-the-land proponent Carla Emery's book *The Encyclopedia of Country Living* (still in print and up-to-date today). In addition, she attended a cheesemaking course at the University of Guelph, in Ontario, Canada (a highly respected program that has been offered since 1956).

Alyce says that one of her best teachers was "lots of trial and error." Even after twenty-one years of success—including numerous American Cheese Society first-place awards—Alyce continues to seek out educational opportunities: She has taken courses at Cal Poly Pomona, Washington State University, and the Vermont Institute for Artisan Cheese (VIAC), as well as attending American Cheese Society conferences and educational sessions.

and communing with other cheesemakers are all viable routes for continuing education. Keeping your knowledge expanding and your awareness of the process growing will help to ensure the quality of your products as well as your own personal and professional gratification.

Being a part of a growing culinary tradition is exciting! Thanks to the perseverance of a handful of America's original artisan cheesemaking companies, the groundbreaking forays of the cheesemaker pioneers of the artisan revival, and the increased awareness and admiration of the life of the small farmer, it is now easier than ever to build a thriving farmstead cheese business. As you read the following chapters, however, you will learn that it will take far more than excitement and awareness to ensure that your cheese business brings you both financial and personal success.

· 2 ·

Making It Official

Now that you have read about what a prime time it is to make and sell handcrafted cheese from the farm, you are probably feeling excited and just a bit mystified. After all, enthusiasm is one thing, but where do you go from there?

Before we get into the details of business and building design, let's talk about some preliminary considerations, such as zoning; working with the dairy inspector; choosing building plans, getting building permits, and working with a building contractor; and, finally, reviewing a primer on the often daunting Concentrated Animal Feeding Operation (CAFO) permit.

Zoning

One of the first things people who are considering building a small creamery should do is find out if their property and land can be approved for the building and operating of a dairy. Every state and local jurisdiction will have land-use regulations and zoning laws that will determine the preliminary feasibility of your project. The laws governing zoning can vary greatly even within the same state. It is important to contact the correct authorities early in the process and receive a reliable judgment regarding your property and your proposed business.

Who to Contact

Before making any phone calls, you will need to have a legal identification of your property. In some jurisdictions your address alone is sufficient. If you also have a property tax identification number (from your property tax bill) or a plate/lot number, this will be helpful. Once armed with your property's legal identification, you will be ready to call the governing agency for the area in which you live. For most of us, that will be the county government, but some states, such as Rhode Island and Connecticut, do not employ a county government setup. If your county does not list a zoning department, then talk to whoever issues building permits and you should be directed to the proper authority.

What to Ask

Once you are connected to the right department, give them your name, identify yourself as the owner of the land (unless you lease or rent, of course), and ask if the land-use laws allow for the building and operation of a small dairy and cheese-making facility. Find out if there are any specific restrictions, such as the number of animals you can own. If you live outside of city limits but not far from a town, find out where your town's long-term urban growth boundary lies. While you may live in a seemingly rural area now, it may be slated for future growth, and even if that growth area does not encompass your land, it could negatively impact your future farming activities.

Important questions regarding zoning:

1. Is my land approved for agricultural uses?
2. If so, how many milking animals of my chosen species will be allowed?
3. Are there any urban growth boundaries nearby?
4. If a farming operation is allowed, is it also acceptable to sell farm products directly from the farm?
5. If the land is not currently approved for farm use, is it possible to apply for a variance or zone change?

Once you confirm that your property is zoned for a small dairy and cheese facility you are ready to make the next call, to the dairy inspector.

The Dairy Inspector

You might wonder why you would be calling an inspector when you have nothing to inspect. But in a sense you do—your plan. By establishing contact with the proper regulatory agents early in the process, you will be doing two important things: First, you will be developing a good, up-front working relationship with an individual who will play a large role in your future professional life. Second, by discussing your plans before you invest any money in architectural drawings, permits, etc., you will have the opportunity to make choices that could prevent costly changes later in the process.

Who to Contact

Finding out who to contact can be a bit of a challenge in some states. For most, your state's department of agriculture will license and inspect dairy facilities. Try searching the Internet with the following terms: "*[your-state's-name]* department agriculture dairy" or "*[your-state's-name]* dairy inspection." By following links, you will most likely find the right department. Or, if you happen to have another dairy, cheese plant, or creamery in your area, you can try contacting them for the information. Regional cheese guilds can be of assistance also. (See appendix A for a list of cheese guilds.)

What to Ask

Important questions for the dairy inspector:

1. Does the state have any instructional materials or guidelines?
2. At what point in the process does the inspector want to become involved?
3. Are there any other resources that the inspector can recommend?

Hopefully your first contact with the inspector will be positive. Remember, many inspectors may not seem encouraging (that's not their job), but most are helpful and supportive. It is their job to enforce the regulations. If particular rules in your state do not seem to make sense to you, don't blame the messenger! The more professional and positive you can make the relationship, the more it will benefit you in the long term.

Building Plans, Permits, and Contractors

Once you decide on a floor plan (ideally after reading part 3 of this book and studying the floor plans in appendix B) you will need a set of building plans drawn by a qualified person. A qualified person does not necessarily mean an architect, but it must be a person familiar with the building codes that will be enforced in your local jurisdiction.

Once drawn to your satisfaction, the plans will be submitted to the building or permitting department for approval. There is often quite a long period during which changes to the building plans might be required; also, permits (such as for septic and water) need to be approved prior to approval of the building plans.

Thanks to some experience I had drawing up plans and getting permits for smaller projects, such as additions and remodels, I was able to draw the full set of plans for our creamery. There were minor changes made by the county and a few by our builder. But, for the most part, it was a sound plan and a great—if challenging—experience.

Factors Affecting Plan Approval

1. Proof that zoning requirements have been met and are acceptable.
2. Engineering requirements, load calculations, etc. (a competent architect should be aware of these requirements and include them in your plan).
3. Infrastructure, such as septic and water (dairy waste will be discussed later in this chapter).

       a. Do current systems meet the standards that will be required for your project?
       b. If no systems exist, permitting and completion of an approved septic system will most often be required.
    4. Fire Department approval.
       a. Setbacks and abatement boundaries for rural settings.
       b. Plan review for fire safety.

Permits

Most states have an agriculture building permit exemption. Your building or some of your buildings may qualify for this exemption. It is best to clarify the parameters that your jurisdiction enforces. If you are using licensed building contractors, the permits will be issued to them, but make sure that they are aware of possible agricultural exemptions. Our barn and creamery did not qualify for the ag-exemption due to its mixed-use nature. But the permit fees were not high in comparison to those for a residential structure.

If you are planning on selling cheese directly from your farm, try not to use the terms "store," "retail," or "commercial" in your plans. For construction purposes, these terms can send up a red flag that might cause building departments to treat your plan as a "commercial building." Even though you will have a business, an agricultural enterprise is not technically commercial with respect to its purpose—commercial construction refers more to industrial and retail complexes, restaurants, apartment buildings, etc. Farmers are often allowed to sell what they produce directly from their farm without extra licenses and

Pholia Farm's dairy barn and creamery during construction.

permits. If your jurisdiction requires an additional business license to sell products at your farm, that can be addressed at a later date; this should not influence your building permit.

Building Contractors

Don't wait until you are ready to build to start looking for a contractor. The good ones will likely be booked far in advance. Talk to friends and get referrals whenever possible. If you see a barn or building project that you admire, try to find out who the contractor was. We all have heard the stories of projects that are far over budget and off-schedule. While this is not always avoidable, taking your time and finding a qualified contractor who is well recommended can save you both money and anxiety.

In most jurisdictions you can serve as your own building contractor (also known as owner-builder). *This does not mean that you have to do all of the work yourself!* You can hire people to do the plumbing, electrical, building, etc. But you will oversee the project and be responsible for the work. If you are the one who "pulls" (requests) the permit, *you* are the one who will be held accountable for failed inspections.

To assist the contractor in completing your project on time and within budget (or at least close to it), follow these tips:

1. Do not make floor plan changes during construction.
2. Know ahead of time what surfaces, colors, fixtures, etc., you want to use, then stick with them.
3. Be on the site and involved—but not in the way!
4. Offer to make hardware store runs, pick up materials, clean up the site—anything that can keep the high-paid labor doing their specialized work on the site.

CAFO and the Dairy Wastewater Management Permit

A dairy and creamery will create wastewater that must be processed and disposed of correctly in order to not overload the land and any watershed areas with chemicals and waste. You will be dealing with three main types of wastewater: animal wastewater, wash water, and human waste from any bathrooms. In addition, whey from cheesemaking must be handled as a pollutant and disposed of in an approved manner. We will discuss the infrastructure options for dealing with wastewater and effluent more in part 3, chapter 7, "Infrastructure and Efficiency," but for now let's talk about the permitting surrounding dealing with dairy wastewater.

CAFO stands for Concentrated Animal Feeding Operation. The Environmental Protection Agency (EPA) designed these permits to protect water supplies from animal waste contamination by very large, confinement livestock operations.

CAFOs are designated by size from small to large. An example of a small CAFO is 300 or fewer cows (the number limit varies by species). In most states, if you have just a few family farm animals, you are not required to obtain a CAFO permit. But if you have a licensed dairy *of any size* you may be required to obtain a CAFO permit. (The term "CAFO" has such a negative connotation for many people, conjuring up images of factory farms that house hundreds, thousands, or even tens of thousands of animals in a confined space, often with limited access to fresh air, exercise, and sunshine. It would be nice if the terminology associated with this type of permit evolved to better describe extremely small operations with only a handful of animals that have year-round access to pasture.)

> **WASTEWATER MANAGEMENT: HUMAN VERSUS ANIMAL SYSTEMS**
>
> Our local sanitation department allowed us to process our dairy wastewater using a very large septic system that is also connected to a bathroom. In this case, since the state does not have anything to do with permitting systems that process human waste, the state CAFO folks deferred to the local regulators. If, however, this same septic system did not have a bathroom attached to it, then it would be an animal waste system and a CAFO permit would be required. We didn't find this out, though, until after I had written an entire AWMP (Animal Waste Management Plan) and filed for the permit! I did learn quite a bit from the process, though.

A part of the CAFO permitting process is the development of an Animal Waste Management Plan (AWMP). The AWMP calculates the volume and nutrient content of all waste produced (including dead animals) and designates how it will be handled (whether spread on crops, composted, sold, etc.). It sounds overwhelming, but boilerplates and guidelines are provided and assistance is readily available, should you be required to file for a CAFO permit.

There is an initial fee and an annual renewal fee.

Who to Contact

Your best choice is to contact a representative at the state level. You can usually locate this person by linking from the EPA's website or by searching "*[your-state's-name]* CAFO." The department handling CAFO issues might be water quality, agriculture, or some other department. Your dairy inspector may know who to contact, but he/she will not generally be a part of the permitting agency.

What to Ask

Important questions regarding CAFO permits:

1. How does my state define CAFOs?
2. What animal wastewater options are allowed in my state?
3. Who is the local agent I should talk to?
4. If my local jurisdiction allows for the wastewater to be handled by a dual-purpose residential septic system, will a CAFO permit still be required?

The CAFO permitting part of the licensing process is one of the most confusing and frustrating "hoops" you will have to jump through. Even within the same state you might find some small creameries that are required to file and others that are not. I recommend starting at the top level when learning the rules for your state—don't rely too much on word-of-mouth when trying to find out whether you will need a CAFO permit or not. Try not to be intimidated by the process. You should receive quite a lot of help, and it is educational to learn how the waste from your animals and processing will affect the land and watersheds; in fact you can think of a CAFO permit as proof of your good stewardship of your land and waste management.

If you are feeling a bit exhausted at this point, just thinking about all of these permits, plans, and processes, you are not alone! But this is just the beginning of the preliminary work. I think of it as an endurance test: If you can make it through all of the initial investigation and education without losing your enthusiasm for the final goal, then you might just be cut out to be a farmstead cheesemaker—and a successful one!

PART II
GETTING DOWN TO BUSINESS

· 3 ·

Sizing Up the Market

No matter how much you adore your animals and how enamored you are of the idea of being a farmstead cheesemaker, your endeavor won't have much chance of surviving—much less thriving—if you don't approach it first and foremost as a business. Asking some important questions now will help you find your niche early on and increase your odds of success: *Who will buy your cheese? What cheese will you make? Where will you sell it? How much will you charge? How will you ensure your place in the market?* These are the questions that should be answered well before you start sketching floor plans or buying equipment. Take time to research and answer these questions. The information you gather will help you develop a sound business plan (more on that in chapter 4, "Your Business Plan and Company Structure").

Market Research for Cheesemakers: The "Five Ps"

When working on your marketing plan, use the "five Ps" of marketing: *People, Product, Place, Price,* and *Promotion.* Let's talk about assessing your market potential using each of these points. Below, I have defined the five Ps as they pertain to the artisan, farmstead cheesemaker, and I've included pointers on how to answer these questions when doing your market research. During all phases of this foundation work, try to remember that all of the answers must come together in a way that leaves no doubt as to the feasibility of your proposed business.

People
Who will buy your cheese? The local population, tourists, and/or people outside your area? Here are some things to consider when sizing up the potential consumers of your products.

Local Population
The first choice for a "green" customer base is the local population. Here are some questions that will help you assess this market:

1. *What cheeses are available locally?* Check your local markets, grocery stores, and cheese counters. Ask the staff at each of these which cheeses are popular sellers, and what kinds they would like to see produced locally.
2. *What percentage of the local market is dominated by staple commodity cheeses (such as most Cheddar, mozzarella, and Monterey Jack) and what proportion is "gourmet"?* While many stores carry specialty, artisan, and import cheeses, it is a good idea to find out what proportion of the market they actually occupy.
3. *Is the local market expanding or static?* Your market might be just beginning to value handmade cheese through an expanding awareness of food quality or growing affluence. Assess this through feedback from store clerks, chefs, etc. If there is demand but no

Terry Carlson, Fairview Farm, Oregon, manning a farmers' market booth. Notice the large banner and photos of the farm's goats and creamery.

supply, and you feel you can meet that need, then this is a market area you can tap into.

Tourists

A year-round or seasonal tourist economy can greatly augment the sales of your cheese. Here are some questions that will help you evaluate this type of market:

1. *Is the tourist market seasonal or year-round?* If you are going to milk seasonally and are planning on making fresh cheeses, it is important to see how the tourist market will mesh with this type of production plan. You might be able to time your production to better coincide with the demand. For example, if you live in an area where the tourist season occurs in the winter, say a ski resort area, but your animals are not in milk during the winter, consider producing an aged cheese that will be ready to sell during that time.
2. *In what venues will tourists buy and/or consume high-end cheese?* Include restaurants, farmers' markets, specialty stores, and farm stands in this survey. Talk to chefs and buyers to find out which products are in short supply and if they would be interested in working with a small local vendor. (A note about chefs: While many top-quality chefs will want to work with your cheese and you, many of these relationships are short lived, as the reality of ordering from many small vendors can prove difficult and time consuming for many chefs.)

Non-Local

Perhaps your largest customer base will be outside of your region or state. This can be the hardest market to tap into. If your research tells you that your sales lie farther afield, you will need to determine ways to make this market aware of your product (see sidebar). Options include:

1. *Cheese competitions*—Doing well can instantly increase your exposure, as many retailers and distributors attend competitions seeking out new cheeses.
2. *Sending samples to buyers*—Retailers often have specific schedules for tasting new products and determining their place in their stores.
3. *Trade show events*—Events such as the American Cheese Society's "Meet the Cheesemaker," cheesemaker receptions at food shows, and cheese festivals are just a few examples of opportunities to present your product to both the public and industry representatives.

> ### PHOLIA FARM: PEOPLE AND PLACE
>
> We were very lucky to have our market find us. Before we were even licensed, we entered a national competition that had an amateur category. The judges were highly respected, nationally known cheese professionals. When our cheese did well (winning Best in Show in the amateur division), one of the judges, also an author of several books on cheese and a buyer for one of the most highly respected cheese counters in the United States, became a fan and supporter of our farm and cheeses. . . . This was a defining experience for us on many points of our marketing plan.

Product

What cheese will you make? Here are some questions to help you focus in on the best product for your business:

1. *What cheese can you create of the best quality and uniformity?* Research and development of your product should occur before opening, if at all possible. While we should all continue to work on improving quality and even on developing new cheeses, make sure you have a product or combination of products that are market-ready by the time you open.
2. *Is there a customer base?* If the cheese you want to make has no market, consider a different choice of product, or at least be prepared with promotional tactics to develop a market for your cheese.
3. *How will you package it?* Packaging can add unforeseen costs and labor. Consider this obstacle ahead of time. For example, most fresh, soft cheeses must be packaged in single-use containers. This adds cost as well as a possible environmental negative for some people (both customers and producers). Try to assess your options ahead of time to make sure your product packaging fits your company's goals.

Place

Where will you sell your cheese? Some ideas include on-farm sales, direct sales, farmers' markets, and wholesale (retailers, distributors, and restaurants). Here are some things to consider when analyzing place.

1. *On-farm sales*—Having a farm stand or farm store offers opportunities to sell your cheese directly to the customer at retail price. (Chapter 13 will discuss more value-added options that can be explored with on-farm sales.) Increased traffic for your neighborhood, liability on your insurance policy, and labor costs are all factors that should be taken into consideration.

2. *Direct sales*—Selling directly to the customer through phone or online sales also allows you to sell at retail pricing. You will have paperwork, possibly the need to accept credit cards, and packing and shipping considerations.
3. *Farmers' markets*—Farmers' markets are another place where you can reap the retail price benefit. Direct contact with consumers can also be rewarding. Travel costs and time spent away from the farm and dairy need to be factored into the equation. Those with nearby, popular farmers' markets often make the majority of their sales through this venue. (For more details on farmers' markets, see chapter 13, "Increasing Your Bottom Line.")
4. *Wholesale*—Wholesale sales will give you the lowest profit margin, but possibly a better return for your time. (When you ship larger amounts of cheese to one buyer, you greatly reduce paperwork and packaging time.) There is also not the waste and loss that can occur due to cut and packaged product that doesn't sell, and samples given out to customers. Here are some possible wholesale customers:
    a. *Full-service cheese counter*—These retailers purchase full wheels direct from the farm or from distributors. They sell cheese cut to order, meaning that their customers have the opportunity to

Cheese counter, Pastoral Artisan Cheese, Chicago, Illinois.

> ### SWEET HOME FARM: PLACE
>
> When Alyce Birchenough and Doug Wolbert of Sweet Home Farm in Alabama first built their creamery in the mid-1980s, they thought they would be shipping their cheeses to retailers. But after experiments with different packaging (all in an attempt to keep the cheese cool while not producing too much waste) failed, they refocused on the local market. Given their location—ten miles from the Gulf of Mexico and a year-round tourist area, as well as in an established "farm-stand" culture—they are able to sell 15,000 pounds of raw-Guernsey-milk cheeses every year almost exclusively from their farm store, which is open two days a week.

      sample the product first and then order and purchase it in the amount they choose. Full-service cheese counters are usually looking for local or regional handmade cheeses. A full-service counter will offer the most personal customer service, and therefore is the best market for unique and high-end cheeses.

  b. *Specialty retailer*—These high-end grocery stores and gourmet markets sometimes sell cheese cut to order, but most often they pre-package it for deli or dairy case sales. Even if retailers are part of a chain, they will often feature local products at regional stores and work directly with the producer.

  c. *Chain grocery store*—These usually purchase through distributors, but in smaller regions they might work directly with producers. Probably the least personalized customer service of all the retail venues—and likely the lowest profit margin for the producer.

  d. *Distributor*—Distributors purchase larger volumes and resell to retailers and restaurants. This brings opportunities for larger producers to get their product out to many more venues for the best return. The distributor price is usually less than regular wholesale (more on that in a moment). Working with a distributor eliminates multiple shipments and billing. Once you deal with a distributor, however, the quality control—handling, storage, packaging, condition at delivery, etc.—is in their hands, not yours.

  e. *Restaurant*—Chefs usually place smaller orders and are often not as consistent as retailers in the amounts they purchase. They can increase exposure for your product, though, often resulting in additional sales at the retail level.

Price

How much will you charge? Pricing cheese is a tricky and often confusing issue. You might think you can figure out how much it costs to produce and then charge enough to cover your costs (labor included). But artisan cheese is more likely to be priced according to *what the market will bear.* Sometimes this is enough to cover your costs, and sometimes it isn't. If you do good market research and complete a viable business plan, you should be able to determine how much product and at what price you will need to sell it in order to create a sustainable business.

Wholesale Pricing
When you sell your cheese wholesale the retailer needs to be able to mark it up close to 100 percent, as well as recover any shipping costs. For example: Say I sold 30 pounds of cheese to a retailer at $12.00 a pound with shipping and packaging costs at $42.00. When the retailer receives the order, he will compare the invoice to the contents and make sure he has received the weight he is being charged for. He then adds the costs together for a total cost of $402.00. This is divided by the pounds of cheese. He finds that the cheese costs him $13.40/lb. This retailer doesn't do a full 100 percent markup, but 95 percent. He multiplies the cost by a factor of 1.95 (if it were a 100 percent markup, the factor would be 2.00), to find that he must charge $26.13/lb. Here is the formula:

> **WHOLESALE VERSUS RETAIL PRICE**
>
> I have often heard cheesemakers say, "I charge one price to everyone; if they want the cheese, they will just have to deal with it!" This may work if you don't have any retailers too nearby, but retailers need to mark up products to as close to 100 percent as possible to help them cover overhead. You could harm business either by having your wholesale cost too high or by undercutting the retailer. Figure out a wholesale price you can live with, as well as your own "retail" price for farmers' markets and on-farm sales.

Formula for 95 Percent Markup by Retailer
Wholesale + Shipping/Handling = Cost
Cost × 1.95 ÷ Pounds of Cheese = Retail Price

Note: If a retailer really wants to sell your cheese, he might pay your price and then mark it up to whatever he feels is reasonable. Remember, though, that if another producer comes along with a similar cheese that he can mark up for a full 100 percent, then your relationship with him will come to an end!

Farm Retail Pricing
When selling direct to the customer, whether from the farm, via online sales, or at a farmers' market, it is important to set a retail price that does not undercut other retailers' pricing. This does not mean you have to sell your product with the full 100 percent markup, since you will not have the overhead that a store does. But set a price that reflects an on-farm or direct purchase discount. Here is an example using the same cheese as in the above wholesale pricing guideline: Let's say our cheese is selling in a local retail store for $24.00 a pound (this shop doesn't add in any shipping costs, as it picks up the cheese from the farm). We decide to have an open house and sell cheese directly to the customer. We mark the cheese up by multiplying the wholesale price of $12.00/lb by a factor of 1.75 and get an

LEGAL STANDARDS

Federal Standards of Identity: The United States Code of Federal Regulations contains a list of just under 100 cheese and "related cheese products" whose manufacture and description are legally defined. What does this mean to you? If you are making and selling a cheese listed in the standards, then it must comply with the guidelines stated therein—especially if the cheese will have interstate sales. For example: According to the standards, Colby cheese must be made from cow's milk and Roquefort cheese from sheep's milk. You are perfectly legal in describing your cheese as a Colby "style" cheese, however. If you are uncertain, your inspector should be able to help, but you can also refer to the list at Cornell University's website: www.milkfacts. info; follow the link for "dairy processing."

on-farm retail price of $21.00/lb (this is about a 10 percent discount over regular retail). This is still a nice markup for you, but it will not alienate your retailers. Remember, they have far more costs involved in the actual sale of the product than you do.

Pholia's Formula for On-Farm Retail Pricing (75 Percent Markup)
Wholesale × 1.75 = On-Farm Retail Price

Promotion

How will you build and retain your place in the market? Select a business name and names for your cheeses that reflect your company's desired image. (See "What's in a Name?" later in this chapter.)

Packaging and Labeling

It is important to know that there are different labeling requirements for your cheese depending upon whether you are selling it prepackaged or "to go." If you package cheese for a deli case, then your state may require that specific information appear on the label, such as your manufacturing plant number, ingredients, weight (in ounces and grams), address, and product standard of identity, if applicable (see "Legal Standards" sidebar). If you are cutting or scooping cheese as it is ordered (think of it like ordering a hamburger at a drive-through restaurant), you can simply wrap it in paper or put it in a carton and sell it; no labeling is required. Whichever way you sell your cheese, you will probably want some kind of label. Try to choose graphics, colors,

A beautifully designed cheese label gives both the required facts and a bit of the farm's story.

fonts, etc., that help develop your "brand identity" (see "Brand Identity" sidebar). If you have several different products, try to keep a similar theme for each one. It might seem like fun to have different labels for each (especially if you are creative and have a lot of good ideas), but it is more helpful for building your customer base and brand identity to have continuity in your labeling so customers can recognize your products at a glance.

> **BRAND IDENTITY**
>
> What exactly is "brand identity"? We usually think of brands in regard to large companies and their many product lines, but branding is also important to the small company, albeit with a much different scope. Think of your branding in terms of continuity of a look—a "visual identity"—that reflects your quality and individuality. Use it as an opportunity to share your "story." Remember, with handmade cheese, people are buying a bit of that story, not just a great food.

Brochures, Newsletters, and "Take-Aways"
Create farm and product literature and narratives that tell your story. Information about your cheeses, your farming practices, yourself, and other details that allow the customer to share in the motivation behind your business will help build loyalty and increase customers' positive experience. A brochure should contain information that doesn't change too frequently. People often keep brochures for extended periods of time and might refer to them in the future, expecting the information they contain to still be pertinent. For information that changes frequently, such as hours and events, either make sure to specify dates or include an insert or separate flyer for transient information. An online or emailed newsletter can also be a good way to engage customers and retailers with events and happenings at your farm.

Media Coverage
Sending updates to local news outlets with farm and cheese news can lead to very useful media coverage. (Develop an email group address book so you can easily send your press releases to all the local media outlets at one time.) Writing stories about your farm and your cheese for interested publications can also increase exposure and enhance your reputation.

Website
Having a website can be another useful way of sharing your story. A website can be purely informational or promotional, or it can also be used for online sales. I advise learning how to run your own site, if at all possible. It is the best way to stay on top of updates and changes to the information without having to pay someone for even minor alterations. If you are not interested in running your own website but you still want one, devote a percentage of your monthly budget to maintenance fees. Remember, a poor website is worse than no website.

Paid Advertising

For most small, artisan cheese companies, paid advertising is not necessary for sales. It can, however, help cement your standing in some valuable circles, whether professional or geographical. Advertising is not only about sales; it is just as much about image. Paid advertising can incorporate such avenues as sponsorship of worthy charities and support for events and organizations, which in their turn will support your goals as a cheesemaking company.

Competitions

Consider entering cheese competitions to build a reputation and customer pride, as well as to get professional feedback regarding your products. Many buyers attend well-known competitions to seek out new, unique cheeses and small farms.

What's in a Name?

Choosing a name for your company is an important step—and rather fun, too! The name should reflect your company goals, your farm's personality, and your product image, and it should have the ring of truth. Here are some extra pointers and questions to help guide you in choosing a really great name for your cheese business:

- **Company Goals:** Do you plan on staying small? How about being family operated? Do you want to create a graphic image in people's minds when they hear the name? Do you want to link your company name to a regional or cultural heritage?
- **Product Image:** If you want to name your creamery "Bill's Cheese," you probably won't want to give your cheese names like "Antoinette's Bliss" or "Sunrise Serendipity." By the same token, if you are drawn to tongue-in-cheek names for your cheeses, such as "Porcupine Pyramid" or "Tubby Tomme," then a creamery name that is erudite might not be the best choice.

CERTIFICATION LABELS

From "certified organic" to "salmon safe," the number of certifications that the small creamery can seek continues to grow. The usefulness of a given certification label varies greatly, depending upon the public's perception of the label's importance and the "dilution factor"—when too many labels accompany a product, it tends to reduce each one's individual importance. Use of the same label by large corporations that produce a similar product also tends to reduce the value of the label for the small producer. You will need to weigh the cost of the certification versus its benefit. If your market is mostly local and the consumer can be educated about your farm practices, it may be that no further "certification" will be necessary. If there is a particular issue that you feel especially passionate about, however, you may want to obtain official recognition for your efforts in that regard.

REAL EXAMPLES OF DESCRIPTIVE TERMS

Kennebec **Cheesery**
Joe's **Dairy**
Kenny's **Farmhouse Cheese**
LoveTree **Farmstead Cheese**
Black Mesa **Ranch**

Ballard Family **Dairy and Cheese**
River's Edge **Chèvre**
Matos Cheese **Factory**
Jasper Hill **Farm**

Silvery Moon **Creamery**
Monroe Cheese **Studio**
Mama Terra **Micro Creamery**

- **The Ring of Truth:** While consumers have grown accustomed to a certain level of "spin" in the advertising of big business, they will expect better from the small, artisan producer. If you name your company "Smith Family Dairy" or "Green Pasture Goat Cheese," your consumers and retailers will be disillusioned if they don't find a couple of Smith children helping with the milking or if your goats are confined to dry-lot paddocks.

> **While consumers have grown accustomed to a certain level of "spin" in the advertising of big business, they will expect better from the small, artisan producer.**

How will you define your business? Will you be a "cheese plant," a "creamery," a "fromagerie," or a "cheesery"? *Fromagerie* is the French term for a facility that makes cheese. Here in the United States, you will find the term "cheesery" (a literal translation of the French term) used mostly in the northeastern states that share a border with the Canadian province of Quebec. The term "creamery" was originally applied to plants that collected cream and milk from several small farms in order to churn butter, make ice cream, and produce cheese. Butter production has shifted to larger, more industrial facilities, for the most part, and while "cheesery" is a more accurate term for a plant that produces only cheese, "creamery" has become the more common term. But feel free to name your business a ranch, farm, factory—anything you want! (See the sidebar for some examples of names of a few farmstead producers in the United States.)

Don't get too bogged down with this decision, but do spend a bit of time coming up with a name you can live with for the life span of your business! Remember, you will have to tack on the legal description of your company (covered in the next chapter), such as LLC or Inc., unless you also file a "doing-business-as" (DBA) name.

A picture should be forming in your mind by now: You should have an idea of who will buy your cheese, what cheese you will make, how much you will be able to sell it for, where it will sell, and how you will promote your product. If you have done this part of the research well, then you are ready to move on to the next stage—the business plan.

While all of this planning and information collecting is not very romantic or exciting, it is the solid foundation that could make the difference between success and failure. Keep up the good work and read on!

· 4 ·

Your Business Plan and Company Structure

For most of us who would rather be outside working with the animals or in the kitchen crafting tasty cheese, the thought of having to sit behind a desk and write a "business plan" brings a kind of glaze to the eyes and numbing to the brain. But try to think of it as an expression of your vision, creativity, and unique ideas for your future—presented in a way that will help you envision your future as a cheesemaker, as well as possibly help you qualify for financial assistance. Even the smallest, best-funded business will benefit from taking the time to create a simple business plan.

This chapter is designed to motivate you to create a business plan that will help you solidify your goals and get financing. If needed, there are many additional resources and guides—books, websites, and the Small Business Administration—that provide templates and ideas for creating a good plan.

Now is also the time to think about what sort of formal business structure your company will take; from sole proprietorship to limited liability company (LLC), you will need to make some informed decisions and take some formal steps to create your business's structure. This is another broad topic to which multiple books are devoted, so this chapter will introduce you to some of the more popular options chosen by many farmstead cheesemakers.

What Exactly Is a Business Plan?

I had intended to provide the perfect example of a real cheesemaker's business plan, but as the samples came in from some excellent sources, I saw that they were all vastly different in structure and content. There are just so many alternate ways to create the "perfect" business plan that no single example will tell you what you need to know. So instead I have decided to cover the goals that should be accomplished when writing a business plan and provide illustrative, fictitious examples where needed.

OVERVIEW OF COMMON BUSINESS PLAN SECTIONS

- **Overview/Business Description**
  - Briefly summarize your business, including products, marketing methods, customer base, owners, business structure, etc.
  - Goal: Introduce your business and summarize your strengths.
- **Mission Statement/Executive Summary**
  - Outline the purpose of the business, the values it holds, and the principles that guide its function.
  - Goal: Share the guiding philosophies that make your company a viable business.
- **Vision Statement/Goals**
  - Share your business hopes for the future.
  - Goal: Share achievable long-term plans.
- **Production/Product Information**
  - Describe the product and its production, including labor, equipment, packaging, etc.
  - Goal: Show complete understanding of the manufacture of your product as well as any needs.
- **Market Analysis/Market Demand**
  - Thoroughly explain and describe existing and projected demand, customer demographics, competition, market niche, etc.
  - Goal: Prove that your product has a place in the market.
- **Company Management/Organizational Structure**
  - Provide pertinent biographical information on experience that you will bring to the business. Describe the legal structure the business will take.
  - Goal: Prove experience and competence for this business venture.
- **Financials/Funding/Budgets**
  - List start-up capital required. Give projected budgets for up to the first three years.
  - Goal: List immediate need and show how the company might perform over time.
- **Risk Assessment/Contingency/Exit Strategy**
  - List alternate scenarios and how your company will deal with them.
  - Goal: Show that you have planned for different scenarios, including the failure of your business.

A business plan summarizes all of the important aspects of a business venture. The Small Business Administration accurately describes the business plan as "your company's résumé." Your plan should sell "the bank"—but also *you*—on your company, your product, and most of all yourself! Above is a brief overview of some of the more common titles for the different sections of a business plan, along with a brief description of the content and main objective of each section. Below we will go into a bit more detail for each section, list some questions you should ask yourself when writing the plan, and give a few pointers that might make the process a bit easier. Remember, these titles can be different, the order of the sections can vary, and additional information can be added—there is no single format for a good business plan!

Overview/Business Description

This is your first chance to introduce and impress. Summarize your business in a way that points out the strengths of the company and its product, including market demand and niche. You will expand on each of these areas in the rest of the business plan, so be brief and concise, but cover it all in the overview. *Pointer: While the overview comes first in the final business plan, it is often easier if you actually write it after you've written the rest of your plan.*

Ask yourself the following questions:

1. What type of business are we?
2. What products do/will we create, and what will make them unique?
3. Who will our customers be?
4. Why are we the best people to do this?
5. Where is our company in its growth status (start-up, expanding, etc.)?
6. What is our company's competitive advantage (family farm, organic, etc.)?

Mission Statement/Executive Summary

In Rodney Jones's paper "Building a Business Plan for Your Farm: Important First Steps," he describes the mission statement as "the foundation without which the whole structure can collapse." A strong mission statement should clearly state your guiding purpose, values, and reasons for creating the business. The mission statement is more concrete than the vision statement.

Ask yourself the following questions:

1. Why do we want to create this business?
2. What need is this business filling—personally, in the community, and in the region?
3. How will our social and environmental values impact the business and community?
4. What other values do we bring to this business that will be an asset?

Vision Statement/Goals

Describe your overarching, long-term vision for your business. What sort of goals have you set, and why will those be an asset to you as you grow? Sharing an awareness of the future of your company helps validate your current plan. Make sure that your vision goals do not conflict with goals and values in the mission statement. *Pointer: Sometimes it is easiest to start writing your plan by working on the vision statement first. When the vision of the company is well defined, you can work "backward" to define the details.*

Ask yourself the following questions:

1. Where do you see the business in five, ten, and twenty years? How will it evolve, change, grow?
2. Where do you see yourself in the company as it evolves and changes?

Production/Product Information

Thoroughly describe your product and how it is manufactured, including equipment, labor, and materials used. This section should show your existing knowledge of the process required to make your product, as well as what equipment might be needed to make your business profitable (this will help support any funding requests later in the plan). *Pointer: Even if you are unsure of exactly what cheeses you will make, you must be able to define some sort of production that will provide a sustainable living. You are always free to revise and update the plan as your cheese production changes.*

Ask yourself the following questions:

1. What type of products will we produce, both initially and in the future?
2. Do we already have experience making any of these products?
3. Will other people be involved in the manufacturing?
4. What kind of equipment will be used to process the products?
5. How will the products be packaged, labeled, shipped, etc.?

Market Analysis/Market Demand

Document current and projected market demand for the products you will be making. List any supporting facts for your projections, such as sales figures from existing markets for cheese, cheese consumption data from the USDA or other documentable source, or quotes from media articles on cheese sales and consumption in your area and nationwide. Discuss where your products will be sold (including stores, restaurants, farmers' markets, etc.), specify how much they will sell for (both wholesale and retail pricing), and support the fact that these venues/outlets have a customer base that will buy your cheese. Discuss the competition and why your product will have a niche or edge. (See chapter 3 for more on analyzing the market.)

Ask yourself the following questions:

1. Who will buy my products?
2. Where will I sell my products?
3. Is the market—local, regional, nationwide—expanding?
4. How much will my product sell for?
5. What will I do to promote my products?
6. Why will my product stand out among any competition?

Company Management = Organizational Structure

Specify how the company will be structured (LLC, sole proprietorship, etc.; more on this later in this chapter), who will run the company, what each person's job will be, and why each is qualified to do that job. You can—and should—describe any business or management experience along with farmer-cheesemaker experience that you or your partner have. *Pointer: Don't underestimate the relevance of any employment or work you have done; just try to show how the lessons you have learned from that job will apply to the new business.*

Ask yourself the following questions:

1. What type of legal business structure will our company take?
2. Who will run the company?
3. What will each person's title/job/role be?
4. Why are we qualified to do this work?

Financials/Funding/Budgets

This is probably the most difficult and tiresome part of the business plan to write (unless you love working with numbers)—yet it's arguably the most important section because no matter how promising your company might look in the rest of the plan, if you cannot show a budget that will eventually balance, you really don't have a business that is worth the risk.

The financials section consists of two parts: a list of start-up capital needed, and a budget for the first year. Projected budgets for the following two years are also often required when applying for financing. *Pointer: If you have trouble figuring out the financial section of your business plan, definitely seek help. Local offices for the Small Business Administration and regional economic development offices can be of great help, as can bookkeepers. (The USDA's Farm Services Agency has financial template forms online, but frankly they are rather complicated and intimidating, and more applicable to large-scale operations.)*

> **No matter how promising your company might look in the rest of the plan, if you cannot show a budget that will eventually balance, you really don't have a business that is worth the risk.**

When thinking about start-up costs, don't forget the following:

1. Infrastructure and building costs—remodeling, permits, labor, materials, etc.
2. Equipment needs—vat, pasteurizer, tractor, manure spreader, cheese hoops/molds, etc.
3. Land—lease, rent, mortgage.
4. Operating capital—money to cover such day-to-day costs as phone, Internet, utilities, and even feed for animals until income is established. (Many people overlook this need.)

Remember, you must be able to justify each cost either in other parts of the business plan or as itemized details on the start-up capital list. When you justify an item, describe how it will help your process be more efficient and cost effective. For example, if you want to lease a neighbor's pasture and you have included

> ## BONNIE BLUE FARM
>
> Jim and Gayle Tanner of Bonnie Blue Farm in Waynesboro, Tennessee, shared the following story about their initial budget:
>
> "Budgeted building and equipment exceeded business plan by 200 percent (and that was doing the work ourselves, [with] very little hired labor); insurance was 300 percent more, including product liability; feed and hay 400 percent more; utilities, propane, and electricity only slightly more than budgeted, but if gas for going to farmers' markets and delivering cheese was included the amount would be greater; and cheese supplies, packaging, and labels 200 percent more.
>
> "But we had good news too: Gross sales exceeded expectations by 200 percent our first full year of sales, 2007. And have increased every year including this 'recession' year.
>
> "I would have to say, looking at the numbers now after three and a half years, we were very naive about the true start-up cost. The number tossed around for building and equipping a small dairy is $250,000. We don't feel that is far off. Of course it can be done with much less, but you get what you pay for. Our farm is aesthetically beautiful, and since we entertain farm visitors [with their farm stay/agri-tourism cabins], that is important."
>
> I asked Gayle if they had used their business plan to seek any funding.
>
> She replied, "Yes, we received Tennessee Agriculture Enhancement Program grant dollars to purchase specialized, value-added equipment and help with the cheese cave."

that cost in your start-up capital list, explain why this pasture will help increase your profit margin: "Addition of 20 acres of grazing land will allow the herd size to expand, allowing for increased production to the level that the facility, management, and market can handle while also adding value with the nutritional and sustainable provenance of 'naturally grazed' and 'grass-fed' to the product, increasing its cachet and marketability."

The budget you present should reflect the scenario as laid out in the rest of the business plan. For example, if you have requested funding to lease twenty additional acres for increasing your herd size, then the budget should show both the income from the additional milk/cheese production as well as the cost of the lease. If you cannot show that this increased cost will truly raise your profit margin, why would a bank lend money to you, and, even more important, why would you want to take the risk yourself? This is why a business plan should be created even if you are financing the entire operation without outside help.

> **If you cannot show that this increased cost will truly raise your profit margin, why would a bank lend money to you, and, even more important, why would you want to take the risk yourself?**

Risk Assessment/Contingency/Exit Strategy

While none of us wants to think about our business faltering or failing, a good plan will take this possibility into account. Address how you will handle a change in consumer tastes, prices—anything that could happen that would undermine the current plan and budget. You should also delineate an exit strategy; in other words, show how you will resolve your obligations and such should you cease operation as a business. *Pointer: Think of this almost like a last will and testament— no one likes to think about and plan for their death, but those who do always leave others in a better situation.*

Ask yourself the following questions:

1. If there is a market failure—such as too much competition—and the price falls, how will we handle that?
2. If there is a production failure and we lose product, how will we handle that, and what will we do now to prevent such failures?
3. If all owners of the company decide to cease operations, how will that be handled?

I have visited a couple of successful farmstead creameries where an exit strategy was never considered in the beginning of the business. While the couples have made a success of their careers as cheesemakers, now that they have reached an age where it is evident that they will not want—or be able—to continue this type of work, they are having a difficult time figuring out "how to get out." Most have built their businesses on land they don't want to leave, but it is often their only capital asset. In the absence of a retirement plan, these folks face tough decisions about what to do with the business, how to live without its income, and where to go if they were to sell their farms.

What Is the Best Business Structure?

The topic of legal business structure could occupy an entire book, but the goal here is to share what is working best for other cheesemakers. Without a doubt the number-one choice of cheesemakers I interviewed for this book was the limited liability company, or LLC. Another common form is sole proprietorship; general partnerships exist also. I do not know of any small cheesemaker who has formed a C or S corporation. So let's talk about the three most likely legal business structures for the small cheesemaker.

Sole Proprietorship

This form is the simplest, yet perhaps the least advisable, structure that a business can take. It is usually operated by one person or a married couple (since they can be identified as a single entity on tax forms). The owner is—and this is very important—*personally liable* for all aspects of the business: profits, debts, and judgments. A sole proprietorship is the easiest business structure to dissolve, transfer, and report on tax forms. Its advantage is in the ease of formation and management. Its disadvantage is in the liability, as well as funding options. If a legal judgment finds the business liable for injury, then the owner's personal

> ### WHAT ABOUT A DBA?
>
> DBAs (also known as fictitious names, assumed names, and trade names) are an everyday part of business, often without the consumer even being aware of it. So just what is a DBA? The acronym stands for "doing business as." DBAs allow you to use a name for your business other than the name of the owner(s), in the case of a sole proprietorship or general partnership, or the name registered with the government, in the case of limited liability companies and corporations. While you must use your company's legal name on permits, licenses, etc., a DBA can be used on such things as signage, business cards, packaging, and advertising.
>
> A DBA is obtained usually by filing paperwork with a governmental office, such as the county clerk or secretary of state, but regulations and procedures vary by state. For a complete list and links to each state, go to www.business.gov and follow the links for "register a business."
>
> Once you have a DBA you will be able to open a bank account under that name and deposit checks made out to your fictitious name.

property, such as home and land, can be at risk. From the standpoint of funding, often loans and grants specifically exclude sole proprietorships from consideration. Even if you start out with this business structure, you can "upgrade" at any time.

General Partnership

A general partnership is formed when more than one person (or married couple) wish to go into business together as equals in the sharing of profits, debts, responsibilities, and liabilities. The partnership itself does not pay taxes, but each partner will report his or her share of the losses or profits on his/her personal returns. As with the sole proprietorship, the general partnership is easy to set up and inexpensive to form. In addition, legal judgments may find the owner's personal property fair game.

Limited Liability Company (LLC)

The LLC business structure is the new kid on the block, having existed in its current form only since 1997. The LLC is a hybrid of sorts between the general partnership and a corporation in that there is no personal liability for its members (owners) and its members are taxed as in a partnership and sole proprietorship instead of as a corporation. In addition, not every member has to have the same rights and responsibilities. So if you have someone who can bring assets to the venture, but management decisions will be made by other members, these issues can all be spelled out in the LLC's operating agreement. Sounds pretty good, right?

Forming an LLC is slightly more complicated, in that simple articles of organization must be filed in the state in which the LLC is formed. An LLC operating agreement (in which each member's obligations and rights are delineated) is highly recommended but not required.

Beautiful barn and creamery at Twig Farm, Vermont.

More Help on Business Structure

To be sure you make the best decision for your business, it is a good idea to consult your tax accountant and/or an attorney. Excellent books on forming each type of business structure are available; most include CDs with templates for creating legal business documents. Again, the U.S. government's Small Business

CREATIVE BUSINESS STRUCTURING: WHY SOMETIMES ONE JUST ISN'T ENOUGH!

Several small creameries around the country (ours included) have split their cheesemaking business from their goat business and formed either two LLCs or one LLC and one sole proprietorship. While this does require keeping two sets of books for tax purposes, it has the advantage of separating liability and simplifying the possibility of cessation of one business and not the other. As the cheesemaking side of the business is much more likely to encounter a lawsuit, it makes sense to separate the two. This way, if there is a judgment against the cheesemaking side, only the assets of that LLC will be vulnerable. It also simplifies things should you decide to cease business as a cheesemaker but want to continue with your livestock. This may be another topic to discuss with an attorney and/or your CPA before making any final decisions.

Administration is an up-to-date resource on all things related to forming and operating a small business.

There, you've made it through another hoop on your way to making, and selling, farmstead cheese! Don't forget to revisit your business plan over the years; it might come in handy for applying for grants for expansion or educational purposes, as well as serving as a reminder of just you how far you have come.

· 5 ·

Production Costs and Issues

Of the many obstacles that confront the small farmstead cheesemaker, finding insurance, dealing with labor issues, and facing the reality of product loss are often not considered until they are staring you in the face. This chapter will help you learn about choices and options regarding these three topics that will help you make good, cost-effective decisions for your business.

Insurance for the Farmstead Creamery

Believe it or not, I know some commercial dairies that are operating without any insurance, either for the property or for product liability. Without proper farm and liability insurance, you place your entire investment and future at risk! It takes only one lawsuit or incident (even if you are not at fault) to, as the old cliché goes, lose the farm. Please keep in mind that all of the information presented in this book is intended to be used as a guide, not as legal advice and counsel. You will want to thoroughly investigate the options and requirements particular to your state and situation. The information contained here is subject to change and should not be considered complete.

Farm Insurance

Just what is farm insurance? Most farm policies offer comprehensive coverage that takes into consideration all facets of a small farm, including the reality that you will probably live on the premises. You will work with an agent to analyze your coverage needs, including land and buildings; products (loss and liability); personal and commercial property, including animals and feed; automobile; employee and visitor liability; and medical (as well as anything else that has personal or commercial value and/or risk).

When we were first seeking coverage, we wanted to keep the current policies we had for our automobile coverage. We had been insured by the same company for decades and were always pleased with the service and premium. But we were not able to use that same company to insure the farm, as they did not offer dairy farm

Pholia Farm's washed rind cheese Wimer Winter aging.

coverage. (Many companies will insure horse or hobby farms, but not commercial dairies, no matter how small!) We looked into splitting the policy, with our old insurer covering auto and the new one covering the farm, but it was pointed out to us that, as a family farm and residence, everything we owned could be linked to the business in the event of injury, loss, or liability.

The following are descriptions and examples of some of the possible areas you will need to consider for coverage. Remember, this is a guide only, not legal advice or opinion.

Product Liability

Sometimes called "food insurance," product liability insurance does just what it sounds like: it provides coverage, to the limit of the policy, from damage caused to others by your products. Most retail venues (such as grocery stores, cheese counters, and even some farmers' markets) will require you to have minimum product liability coverage of one million dollars. Even if this is not the case with your retail customers, carrying product liability coverage is a good idea. Here are some points to consider when discussing product liability:

- Annual product sales (estimate in the beginning)
- Level of coverage needed
- Prior claims
- Quality assurance or HACCP plan (see chapter 12)
- Type of cheese (here they will want to know if it is raw-milk cheese)

General Liability

General liability provides coverage for injuries and property damage related to your farm. This includes employees, interns, visitors, and volunteers. You will need to address the following:

- Number of employees
- Interns (include details, such as housing and duties)
- Events you hold where visitors would be present
- Precautions you will take to ensure safety

Commercial Property Coverage

This category will include the following areas:

- Buildings used in your business
- Property, such as cheesemaking and office equipment
- Business income, to cover loss of income as put forth in the policy
- Agricultural equipment (tractors, trailers, etc.)
- Animals, hay, feed

Personal Property Coverage

As in a homeowner's policy, this coverage is for personal residence and structures, furnishings and belongings, and so forth.

Automobile Coverage

Each vehicle you drive will likely be considered commercial, even if you don't plan on using it for business. The reasoning is that if your designated business auto breaks down, you will probably use another vehicle you own. Remember, you are engaged in business even when you take business mail to the post office. The good news is that usually a policy that bundles so many areas together is relatively affordable (as compared to purchasing each component separately).

Finding a Company and an Agent

This can be quite a challenge—especially if the phrase "raw milk" is involved. Even though raw-milk cheeses are legal, they still raise a red flag for many insurers. If you know other cheesemakers in your state or area, you can talk to them about who they use, and whether they are satisfied with them. Some states' agricultural departments maintain a list of companies that provide farm insurance, but many of these providers focus on crop farms and not dairy farms. Agricultural newspapers often have ads as well, but, again, the majority do not necessarily cover farmstead cheese operations. Another very good place to find a company is farm/agricultural expositions. Quite often insurance providers will have a booth at such events.

A local agent is a must, in my opinion. Our first agent, who was some distance away, was recommended by another cheesemaker in our state. I tried to find a local agent at that time who could write policies from the same provider as the long-distance agent, but without success. When we finally located a local agent and switched, our policy became more pertinent (and even a bit less expensive). We also felt more satisfied with the service. Remember, you can have an agent visit your farm before you sign a policy. He or she will be working for you, so take the time to interview and find the right match. The complexity of the whole farm policy requires a good working relationship with the agent and frequent reassessment for updates and changes.

> **The complexity of the whole farm policy requires a good working relationship with the agent and frequent reassessment for updates and changes.**

What Will It Cost?

I wish I could tell you! Everyone's policy will vary greatly. But you can start your estimate at $350 to $550 per month for a small creamery. Remember to reassess your policy on an annual basis. It is possible that reductions can be made the longer you are insured without incident. Here are some pointers:

- Read the policy yourself and look for coverage that is not needed or that is missing.
- Consider higher deductibles for items that you believe are low-risk or that you can afford to cover out of pocket.
- Don't make claims for small-value items; remember, every claim will be taken into consideration when you are renewing your policy.
- Shop around!

Health Insurance

Many small farmers in the U.S. work and run their farms without health insurance. The high cost of individual policies is quite prohibitive. It is hard for many of us to contemplate spending hundreds of dollars each month on something that we may never need—to the extent of its coverage—when everywhere we look on our farms there are real, everyday problems that those same dollars would solve: a new roof on the barn, a new pump for the well, a ton of hay for the animals. But don't fool yourself: uninsured farmers are playing a game of chance. Some will win, but others will lose—and by lose I mean lose everything they have worked so hard to build, including that barn with its new roof.

> **Uninsured farmers are playing a game of chance.**

The best wisdom I've heard on this topic was shared by Marion Pollack and Marjorie Susman of Orb Weaver Farm in New Haven, Vermont. For them, the health insurance bill is always paid first, no matter how tight the budget might seem at the time. They understand that it would take only one incident, even the breaking of a leg, to financially cripple the entire operation. When all of your assets are tied up in one property and one livelihood, risking its loss by not having health insurance is an unwise business decision.

Many farmstead cheesemakers rely upon insurance coverage provided by the employer of a spouse who works off the farm or, as in our case, health-care coverage from military retirement.

At the writing of this book, several hopeful cooperative health insurance plans are in the works, including one through the American Cheese Society. Other organizations in various parts of the country offer members health insurance options as well, including the Farmers' Health Cooperative of Wisconsin, the Agri-Business Council of Oregon (there are branches in many other states, but as far as I can determine, only Oregon's offers a group health insurance plan), and Agri-Services Agency. With health insurance being on the forefront of the political agenda at the moment, it is possible that there will be better access to care for all farmers in the near future. At the moment, though, it is costly and difficult to find.

Life Insurance

As a small business that likely relies on only a few people (usually family members) for all aspects of operation, the loss of one of these key members can mean the loss of the business. Sit down with your business partner(s)—husband, wife, mate, etc.—and do your best to try to realistically calculate the amount of funds needed should one of you die. While of course no one can put a price on replacing a lost loved one, a certain amount of money can help you weather the loss and eventually recover—from a business standpoint—without significant damage to your business. The good news about life insurance is its affordability, especially when compared to health and farm insurance!

When choosing life insurance coverage, ask yourself a very basic question: *"If my partner died, how much money would I need to keep the farm, the business, and the family going?"* Consider the following possible costs:

- Funeral and burial, or alternative life-end, costs.
- Hiring help to replace lost labor (don't forget things like cleaning, cooking, and child-rearing).
- Debt and bills previously paid by the lost family member's financial contribution.
- Lost benefits, such as health insurance, retirement pay, etc.

Even if you believe you would sell your farm and business should your family partner die, you will still have costs associated with the transition that could be difficult to cover.

Good Help Is Hard to Find: Labor Issues

Here comes the topic that brings the most frustration to the very small farmer—finding reliable and competent workers. The nature of the life of a dairy makes it inherently difficult, with milking taking place at 8- to 12-hour intervals and the unpredictability of animal behavior. Of all of the issues facing the small dairy person, finding good help seems to be the most perpetually frustrating. In my interviews, the only people I came across who had not experienced dissatisfaction with farm help were those who didn't use any.

> **The only people I came across who had not experienced dissatisfaction with farm help were those who didn't use any.**

Employees or Interns?

Many small dairy farmers find the concept of an intern very appealing. Who wouldn't? In theory, it's a win-win: trading work experience for labor. But be warned: state and federal labor laws still apply to this relationship, and lawsuits have been brought in the past due to dissatisfaction on the intern's part. This section will cover the basics, but it is not meant to be a complete guide to legally employing and engaging labor of any kind. Be sure to consult with your labor office, lawyer, bookkeeper, or other pertinent and up-to-date resource.

What's the Difference?
While in practice interns are often treated very differently from employees, the law recognizes very little difference. Interns exchange their labor (usually seasonal) for housing, food, and, most importantly, hands-on education. In addition, they are almost always paid a stipend. Employees, on the other hand, exchange their labor for cash wages. In both cases, federal and state labor laws apply. You as the employer could be held responsible, in the eyes of the law, should your workers—employees and interns alike—have a grievance with your employment practices. Remember this and protect yourself, even if at the time it seems like more trouble than it's worth.

> **You as the employer could be held responsible, in the eyes of the law, should your workers—employees and interns alike— have a grievance with your employment practices.**

So given the complications, why would you choose an intern over a traditional employee? It seems that the best intern relationships come when there is an underlying desire to teach and share on the part of the farmer-cheesemaker. When that desire is paired with a worker whose primary motivation is not to earn but to learn, then a mutually beneficial and satisfying relationship can occur.

Legal Considerations
- **Internal Revenue Service rules for reporting compensation and withholding taxes.** Download or send for the "Agricultural Employer's Tax Guide" (www.irs.gov/publications/p51/index.html). You will need to report all compensation (including meals and housing), but you will not have to withhold taxes on the value of room, board, or other non-cash wages. If you employ a bookkeeper, be sure

---

TIPS FOR SUCCESSFUL INTERNSHIPS

1. Clearly define and support (with documentation or other evidence) the monetary value of room and board.
2. Document hours spent on non-agricultural work, such as packaging product and selling at farmers' markets, for accurate tax reporting and compensation.
3. Clearly define educational goals and how they will be attained by the intern.
4. Document time spent on educational instruction and subject(s) covered.
5. Structure stipend or wages in a tiered approach, with a greater educational component in the beginning of the internship and lessening toward the end.
6. Provide social opportunities for solitary interns during their off time.

On most small farms, family members provide the majority, if not all, of the labor. Here Vern Caldwell, the author's husband, feeds young kids at Pholia Farm, Oregon.

he or she has a copy of this guide, as rules for agricultural workers are often unique.

- **Federal minimum wage law.** While most small farms are exempt from the federal minimum wage law, you should review the Fair Labor Standards Act as it applies to agricultural employers. It covers other issues, including overtime rules and "non-agricultural" work that takes place on or off the farm. For example, selling at a farmers' market is *not* considered agricultural work, and the federal minimum wage law would apply. (In 2009 the federal minimum wage was raised to $7.25 per hour.)

> ### FRAGA FARM
>
> Jan Neilson of Fraga Farm in Sweet Home, Oregon, has a very well-thought-out intern and employee program. Her two part-time employees and the occasional intern are all on payroll (done by a bookkeeper once a month, costing Jan only about $25.00). The intern's work is divided into hours spent doing agricultural work (such as milking and feeding) and non-ag work (such as farmers' markets and packaging of product). From the intern's gross pay are deducted the non-cash compensations (such as room, board, and education). Then taxes, Medicare, workers compensation, etc., are withheld. It sounds complicated, but with the help of a good bookkeeper, Jan was able to set up a legal and fair program that allows her to have the workers without the worry.

- **State minimum wage laws.** A few states have minimum wages that are higher than the federal level. In addition, some require that minimum wage be paid to all farm workers. Check with your state labor board.
- **State compensation laws.** States will vary in how they view compensation such as room and board. Some will allow you to set the value of the compensation, while others stipulate an amount.
- **Workers compensation laws.** Go to the U.S. Department of Labor's website (www.dol.gov) and follow links for the Employment Standards Administration (ESA) and the Office of Workers Compensation Program (OWCP) to find a link to your state's office. Again, states vary in their laws, so be sure to investigate thoroughly.
- **Educational component.** If you can thoroughly document that your training program will truly educate and benefit the trainee, this can be a form of non-cash compensation, provided that both parties have agreed on its value.
- **Work agreement.** However you decide to structure your internship or employment program, a mutually agreed upon and signed work/internship agreement is highly recommended—and in most states it is a requirement.

For more thorough information on setting up successful internships, I recommend the following reading:

1. *Internships in Sustainable Farming: A Handbook for Farmers,* by Doug Jones, published by the Northeast Organic Farming Association. Available at http://nofany.org/publications.html.
2. *Western Sustainable Agriculture Research & Education (SARE) Farm Internship Curriculum and Handbook,* by Tom and Maud Powell and Michael Moss. Available at http://attra.ncat.org/intern_handbook.

> **PHOLIA FARM**
>
> Every year we have had something happen that has caused the loss of up to 10 percent of our annual production. One year it was inadequate cooling in the aging room (it didn't cause us to lose cheese, but we had to shut down production for a week to rebuild the aging room); the next year it was cheese mites (these are normal on long-aged cheeses but have to be kept at a minimum or they cause severe aesthetic and flavor defects). Another year we lost several months' worth of cheese to poor quality due to high milk urea nitrogen (MUN) levels in the milk (caused by overfeeding of a certain type of protein). The lesson for us was to be more alert at all stages of the process, to keep learning, and to learn to accept the occasional "glitch" in income.

3. "Agricultural Employer's Tax Guide," published by the Internal Revenue Service. Available at www.irs.gov/publications/p51/index.html.

Having interns can either be a very rewarding experience or it can be a very frustrating one. The cheesemakers I interviewed across the U.S. had experienced the entire gamut, from satisfaction to knowing someone who had been sued by a former intern. The same is true of employees. The best advice is to understand the legal requirements, structure your program to fairly meet both your and the intern's goals, and document that program. And finally, don't have any second thoughts about investing in the services of a good bookkeeper. Do you really have the time, skills, and interest that is required to keep your farm's financial side in order? A few hours' work a month by a competent bookkeeper is usually all it takes to keep a small farm's payroll and books in good shape.

Don't have any second thoughts about investing in the services of a good bookkeeper.

Product Loss

Your first business plan will include income projections based on how much milk you will have and how much cheese you will make from that milk. While in theory these numbers should work, in reality you will likely have a significant amount of waste—from milk lost due to equipment failures; milk quality issues (such as bacterial and somatic cell counts); and lowered production from sick animals. You may also have waste in terms of finished product—from equipment failures, quality issues, expired shelf life, sampling to the public, and even donations to charities.

Along with lost product will often come additional costs related to the reason for the loss, such as replacing and repairing broken equipment (that led to lost milk), treating sick animals (that caused poor milk quality), and product and milk testing (to identify the reason behind the poor milk quality). When creating your business plan, it is a good idea to factor in a percentage of product waste and to budget in a contingency fund for such occasions.

Product sampling at farmers' markets, open houses, and special events can quickly eat through a small chunk of your inventory. For example, if you are providing cheese for an industry event (let's use the American Cheese Society networking breaks as an example), you might calculate ¼ to 1 ounce per person, depending upon the total volume of cheese provided. If 400 people are expected to attend, you might need to contribute anywhere from 6.25 to 25 pounds of cheese. At farmers' markets and open houses, people will probably want to sample each cheese you present. One suggestion is to feature fewer varieties at such events and rotate the types; for example, one week sample out two types and the next week use two different varieties. This can both keep your sample "waste" down and also provide an enticement for shoppers to return to your booth weekly.

Author's daughter Amelia giving out Pholia Farm cheese samples at the local public radio station's annual fund-raising cheese and wine tasting event.

Dealing with requests from charities for product donations can be awkward for the new cheesemaker, and even for some of us who have been at it for a while! I suggest budgeting a certain amount of cheese (either in pounds or percentage of annual production—think of it as a "cheese tithe") to charity. You can either dedicate this to causes you find personally gratifying or choose from among those that contact you. Count on being contacted regularly by many causes, and be ready to explain your policy. Most of these callers will be grateful to not be "strung along" and to know immediately whether or not you will be able to support their event. Here is approximately what we tell new callers: "Your event sounds really worthwhile, but because of our production size we have had to limit our cheese donations to just the few events that we are currently supporting." If you would still like to help them, you can offer to sell them product at wholesale.

From insurance of all kinds, to payroll, to planning for lost product/income, farmstead cheesemakers face many hurdles during the development and growth of their business. Try to remember that these challenges will seem far less daunting after a while. Seek advice and help from industry and regulatory experts as well as other cheesemakers whom you respect, and, most of all, try to keep in mind that being prepared and planning things out for the worst-case scenario will help protect your future as a farmstead creamery.

· 6 ·

Creative Financing

Now that you have a workable business plan, including a budget and a list of needed start-up capital, you will have a good idea of how much money you need to get your cheesemaking operation off the ground. Finding funding is often difficult and can bring added cost in the form of interest rates to your budget. In almost all cases, cheesemakers I interviewed relied on an outside source of income throughout the construction process and even for the first few years of business. In some cases, one member of the team continued to hold an off-farm job or have another source of income, such as retirement or investments. Many cheesemakers sink their entire life savings into the project in addition to accumulating some personal debt. It is actually amazing to see the financial feats we cheesemakers will perform in order to build our "dream." This chapter will help you navigate the myriad of options for funding your dairy.

What's It Going to Cost?

Costs will vary based on several key factors: land ownership, existing infrastructure, outside labor costs, scope of the project, and equipment and supply costs. To construct and equip a creamery of the size and scope covered in this book, the cheapest I have seen is about $15,000 to $20,000. This cost was achieved by owning land with existing buildings and systems, doing all the remodeling and construction work without outside labor, purchasing used equipment, and delaying some improvements until after an income was established. For those who need to build from the ground up (meaning that there are no existing systems, such as power, septic, water, animal housing, etc.), who can

> **TIPS FOR REDUCING FUNDING REQUIREMENTS**
>
> 1. Start shopping for used equipment early in the process: Check restaurant suppliers, classified ads for restaurant auctions, dispersals, etc. Watch agricultural newspaper classifieds as well as Internet auctions and classifieds. Post your "shopping list" to Internet chat groups focused on dairy topics.
> 2. Prioritize projects. Can some improvements or additions be made after you are licensed and have an income from cheese?
> 3. If you have some construction experience, but not enough to head the project, ask your builder if you can be one of the laborers.
> 4. Look for cost-saving construction options—but don't sacrifice quality and longevity to save money; it won't pay off in the long run.

> **MAMA TERRA MICRO CREAMERY**
>
> Armed with a thorough business plan, Robin Clouser and her husband Gabe obtained a loan from a local commercial bank to build their small, family-run creamery. The land already included their home power, water, and septic. They also had existing goat housing, paddocks, and fencing. By doing much of the labor themselves they were able to build their 384-square-foot creamery and dairy for $85,000. The cost included upgraded construction materials, a 30-gallon pasteurizer (which also serves as a bulk tank and cheese vat), a dairy wastewater management system (an aboveground holding tank into which the dairy wastewater is pumped and later spread on fields), and appointments such as sinks and lighting. See appendix B for Mama Terra's floor plan.

do a good part of the labor themselves, and who purchase mostly used equipment, the cost can range between $150,000 and $250,000. You can see that the range is large and greatly dependent upon available resources and infrastructure already on the land, equipment choices, and the owner's skill and resourcefulness.

> **It is actually amazing to see the financial feats we cheesemakers will perform in order to build our "dream."**

Loans

The first place you will probably fill out a loan application will be at a commercial lending company, such as a bank or credit union. If you are turned down, you can seek a loan that is guaranteed by another institution such as the Farm Services Agency (FSA) or the Small Business Administration (SBA). When a loan is "guaranteed" it means that the funds still come from a lending company (such as the same bank that turned you down), but the intermediary company has guaranteed that you are a good risk. Guaranteed loans usually come with a different set of parameters, such as what percentage of the loan the lender is guaranteeing; a guarantee fee that is usually added into the loan; and collateral, often in the form of liens on future products, equipment, etc. Organizations such as the FSA and the SBA usually have a direct loan program as well, with their own set of criteria with respect to interest rate, loan limit, and payback time. For a list of major lending institutions, see appendix A.

If you have decent amount of equity in your land and home—in other words, if the property's value is significantly greater than what you owe the bank—then you can access that value through a home equity line of credit (HELOC) or a home equity loan. With a HELOC the line of credit is used as needed, with interest charged only after the funds are used. With a home equity loan you receive all of the funds in one payment up front and interest starts accruing on the balance

Mama Terra Micro Creamery's small, efficient building contains the parlor, milkhouse, and creamery.

from the beginning. In essence, a HELOC acts like a credit card and a home equity loan acts like a traditional mortgage. Interest rates on HELOC loans are usually not fixed (but often can be at some stage during the life of the loan), whereas interest rates on home equity loans are usually fixed. Repayment times are usually shorter than on first mortgages. An additional benefit to the home equity loan is the potential for a tax deduction on the interest paid—but be sure to check with a tax accountant before claiming such a deduction.

Investors

It is worth exploring some other options for financing your business besides loans (after all, the title of this chapter is "*Creative* Financing"). While investors will bring a new set of responsibilities and commitments, they can also bring capital for funding new equipment and expansion. Family members are a common resource for many new businesses—but be aware of the potential for complications should the business not work out. The beauty of investment from outside is the impersonal nature of the arrangement. If you are considering approaching family members for funding, spend as much time and effort and attention to detail on clearly written, legally binding documents as you would if the investment relationship were with a stranger. By making sure all parties are clear on their obligations, you have a better chance of avoiding confusion and dissension in the future.

Members and Partners

If your business is an LLC or general partnership you have the option, at any time, of adding members/partners who bring capital to the business. These members/partners do not need to have any management power or operations input; they can simply exchange their investment for membership and an agreed-upon amount of the company's profits—such investors are commonly called "silent partners." Be sure to seek legal advice when structuring a partnership with any new members.

Venture and Angel Capital

Venture capital is investment from a large firm, while angel capital comes from an individual or a small group. Venture capital is more commonly associated with investment in existing businesses; angel capital is more common in the funding of start-ups. These types of capital can be in exchange for ownership or follow a traditional repayment-with-interest plan. Both sources are rather unlikely options for the small farmer-cheesemaker, but not out of the realm of possibility. For some resources on seeking venture or angel capital, search www.buzgate.org (America's Small Business Assistance Network); search by state, then search "FundmyBiz."

Community-Supported Agriculture (CSA)

A CSA sells subscriptions to a season's worth of products at the beginning of the season, providing the farmer with working capital for that year. The CSA model is most often used for produce farms; however, it can work for the small dairy farm as well, either in partnership with an existing produce CSA or on its own. A small dairy that will be producing a variety of cheeses and perhaps fluid milk can gain working capital in exchange for guaranteeing a certain volume of products to its subscribers. I feel this can be especially valuable when you are trying to get money to purchase a new piece of equipment. For example: Say you have decided to buy a small milk bottling setup. By selling that milk before it is even bottled, you could recoup a significant portion of the investment price. (See chapter 13, "Increasing Your Bottom Line," for more on milk bottling.) Of course, this works best for established businesses with existing loyal customers.

Grants

Grants are a bit of a pie-in-the-sky solution to funding a new business. Almost all grants provide funding for existing businesses that want to expand, develop new products, add renewable energy, conduct research, or improve business practices (such as add organic certification, focus on local markets, or improve animal welfare). They rarely fund equipment, construction, or personal salaries. For the very large business—one that will provide a significant number of employment opportunities in a state or region that is economically depressed, for instance—start-up grants that do fund equipment and construction are not unheard of.

That being said, grants might still be a route of investment for your farm, especially if you have an existing farm and want to develop a cheesemaking business. Expect lots of paperwork, including business plans, financial statements, feasibility studies, letters of recommendation, and much more. Grants are usually very specific about what they will and will not fund. Before you go to the effort of applying for one, be sure the source it is an appropriate choice.

For help on finding and selecting grants, try your local Cooperative Extension office; the National Sustainable Agriculture Information Service (ATTRA—I know, the acronym doesn't match the name!—www.attra.ncat.org); Sustainable Agriculture Research and Education (SARE; www.sare.org); Grants.gov (www.grants.gov); and USDA's Rural Development department (www.rurdev.usda.gov). The Farm Credit System (mentioned previously under loans) also supplies grants through its FCS Foundation (www.fcsfoundation.org).

Personal Investment

This is the most common source of funding for many small cheesemakers: Whether through savings, investments made in real estate, or pay-as-you-go, many small dairies are funded by their owners/builders. Even though I covered home equity loans and lines of credit earlier when I was talking about loans, these sources of funds really still fall into the category of personal investment. I would not *ever* recommend using credit cards to finance any long-term business needs or costs (anything you cannot pay off the same month you purchased it)—it just isn't worth the cost. While it also may not be recommended to cash in IRAs, stocks, or other investments, some people do.

My favorite choice for personal investment is pay-as-you-go. Of course, this requires some other source of income, as well as time and patience. The benefits of this method include no debt and time to change your mind—on anything from the floor plan to the wall color to the whole idea—as well as time to learn more about the business without the pressure of needing to produce product.

We were very fortunate: We had bought a fixer-upper at the bottom of the Southern California housing market; after we spent six years improving it and then sold the property at the peak of the recovered market, we had capital to start building our dairy. Even so, there was not enough money to complete everything. Consequently, three years later we still don't have our house done—a small travel trailer combined with a kitchen above the barn is good enough for now. We have the best of intentions of finishing the house, but other business-related needs keep taking priority. Someday . . .

When considering your financing options, remember that you can use more than one source—this is where the creativity comes in. Using your business plan and financial projections, try to prioritize each need and find ways to fund the most important requirements first. Often, things that you assume are essential (such as, in our case, a house) need to be delayed in order to get the business off the ground—and then maybe postponed again later, too, to keep the business running well. I see financing a business as an evolving compilation of needs, priorities, and options. There will always be a need; you just have to figure out its priority and then choose the best option.

PART III
DESIGNING THE FARMSTEAD CREAMERY

· 7 ·

Infrastructure and Efficiency

In chapter 2 we discussed some issues related to dairy waste and wastewater management. In this chapter we will talk about choices you can make for handling both wastewater and fresh water, as well as power usage and building efficiency. As costs for power and water will no doubt continue to rise over time, it behooves the small-business cheesemaker to seriously assess these considerations before beginning construction. The more efficiency that can be built in, the better the return over the life of your business—not to mention the good feeling you will get from knowing you are doing your best to reduce your environmental footprint.

Water

Potable (drinkable) water is used in the creamery for the washing of equipment, hands, etc., and sometimes in the actual making of cheese (such as in washed-curd cheeses like Jack and Gouda).

Quality

One of the most important issues to be considered when addressing water supply for your creamery is quality. Your water source needs to be free of contaminants and any elements that could negatively affect the quality and safety of your cheese. Water enters the processing chain in a number of ways, including as residue on hands and equipment; when diluting cheesemaking ingredients, such as rennet and calcium chloride; and when washing curd in such cheeses as Colby and Gouda.

If you receive your water from a municipal water source treated with chlorine, your main concern will be not pathogens in the water but having chlorine-free water for diluting such things as rennet and other coagulants. This can be accomplished by purchasing distilled water or by de-chlorinating—using a small amount of milk added directly to the water (the chlorine molecules will attach to the organic compounds of the milk and be deactivated). You only need to add enough milk to give the water a faint milky color. This method is used in many mid-size creameries where the purchase of distilled water would not be cost effective.

If you have a well or other private, non-treated water source, it is important to have your water tested regularly for coliform bacteria and/or any other contaminants that might be of concern. Be sure to do this testing well before you begin production so you have extra time to use or install any needed remedies, treatments, or equipment and test the water again. Most states require preliminary testing as well as routine testing after you begin production. But even if they don't, you are wise to implement your own testing regimen. In addition, you should expect the dairy inspector to inspect your wellhead and any holding tanks, usually just prior to licensing, although this requirement can vary by jurisdiction.

If your water is not considered safe for manufacturing cheese, then various purification systems can be used. One of the most popular choices is an ultraviolet (UV) purification system. These systems utilize UV light to kill *Escherichia coli (E. coli), Salmonella, Legionella pneumophilia, Mycobacterium tuberculosis,* and *Streptococcus* bacteria, as well as other pathogens. In addition, the same system can be equipped with various filters to eliminate sediment, odors, or other off-flavors. Whole-house filters and reverse osmosis units are also possibilities. It is important to consult with a plumber and water quality expert to determine your needs based on your water's test results and issues. Each filtration system has pros and cons and should be analyzed on an individual basis.

> **Utilization of equipment to its maximum capacity will increase your efficient use of water and lower your production of wastewater per pound of cheese produced.**

One often-unexplored aspect of water quality is how your water works in conjunction with the cleaning chemicals you use. Water hardness (mineral content) and pH (an indicator of acidity or alkalinity) greatly influence the effectiveness of detergents and sanitizers. You should have your water analyzed for pH and hardness; if the levels deviate from average, you will need special detergents. Be sure to consult with the manufacturer of the cleaning products you use, and/or a company representative, to select the proper products.

You should also install an anti-siphon device (also known as a backflow prevention device or valve). In order to maintain potable water quality, you must ensure that water pressure never reverses direction and siphons animal drinking water, wash water, or any other contaminated water back into the water pipes. A backflow prevention device should be installed between the lines running to the creamery and the lines running to the barn. In addition, individual anti-siphon devices should be placed at each hose faucet where the possibility exists that a hose end could be left in standing water (where it could potentially siphon contaminated water back into the main lines). Talk to a plumber or plumbing supply company to choose an appropriate device. Also be sure to consult with your inspector as to placement, as well as any regulatory requirements that might exist.

Calculating Needs

Running a dairy and making cheese requires a lot of water. If your water comes from a municipal source (what we often call "city water"), your increased use will be reflected on your water (or in some cases sewer) bill—and while you might not have to worry about the volume of water available, the likelihood of increasing water costs is something you will need to consider. If you have a private well or other water source, your well pump and any filtration system will be put to greater use and this will be reflected in higher energy bills.

Estimates of water use in the farmstead creamery vary greatly based on cleaning practices and volume of milk processed. The volume of water used goes up in steps based on the potential production capacity of your equipment. For example, if you have a 50-gallon bulk tank and vat but are processing only 20 gallons of milk into cheese, you're really not using that much less water than if you processed 50 gallons—cleanup of the equipment will use the same amount. Utilization of equipment to its maximum capacity will increase your efficient use of water and lower your production of wastewater per pound of cheese produced.

A common ratio used by cheesemakers states that it takes from 1 to 5 gallons of water to process 1 gallon of milk (from the milking parlor to the finished product). Our own personal experience shows that we use about 4 gallons of water for every gallon of milk, by the time we factor in cleaning the parlor, milkhouse, make room, and equipment. The bigger the batch, the fewer gallons of water used per gallon of milk produced.

> **A common ratio used by cheesemakers states that it takes from 1 to 5 gallons of water to process 1 gallon of milk.**

Reducing Use

While water seems like a cheap resource, it should be treated as an endangered one. As human populations grow, the demand on municipal treatment plants and on rural water tables continues to rise. In addition, how a dairy handles its wastewater and how it attempts to conserve and/or reduce waste can become a political tool—to the benefit or detriment (or eventual demise) of the dairy operation. Reducing water use is good from every angle: saving money, saving local resources, and saving your business's image.

In issue 32 of her quarterly newsletter *CreamLine,* publisher Vicki Dunaway quotes Kent Rausch and G. Morgan Powell, authors of *Dairy Processing Methods to Reduce Water Use and Liquid Waste:* "Stop using your water hose as a broom."

While boilers are not commonly used on very small farms, when an economical fuel sources is available they are quite viable options. This wood-fired boiler with heat storage tank (on right) works in combination with a propane on-demand water heater (in background) to provide hot water at Pholia Farm, Oregon.

The best way to reduce use is to employ a squeegee or broom to assist with cleanup instead of spraying every little piece of curd to move it to the drain. Combine this with efficient sizing of equipment and batches and you will have maximized your water use.

Hot Water

You will need a reliable supply of hot water, at about 140–160°F, for cleaning dairy equipment (more on cleaning in chapter 8) and heating the cheesemaking vat. Depending upon your water type (hard or soft) and the type of detergents used, your optimal hot water temperature will vary, but by planning on having it available toward the high end of the range, you will be better prepared for all situations. This seems to be the area where most creameries initially fall short, ourselves included!

Calculating Needs

Base your need calculations on the high estimate for total water usage (mentioned above) and calculate that two-thirds of that will likely be hot water. Remember that if you are sharing a hot water source for domestic use as well, you will need to consider not only volume but also pressure. Ask yourself, "If hot water is running in another part of the building, will I still have enough for cleaning in the creamery?"

TABLE 7-1: Comparison of Water Heating Equipment Options					
Type	**Gal/Min**	**Volume**	**Consistency**	**Initial Cost**	**Operating Cost**
Standard Storage WH	Based on water pressure	Limited to volume of tank	Temperature drops as tank is emptied of hot water	Lowest	High; cost can rise yearly based on power costs
Tankless On-Demand WH	2–5 gal/min*	Limitless	Consistent	Higher than Standard	Better than Standard; cost can rise yearly based on energy source
Heat Pump WH	Much more efficient electric choice than standard WHs; pulls heat from surrounding air to heat the water; works best if installed in a warm room or space; will vent cool air			Higher than Standard	Improved, dependent upon other factors; payback time can be only about 5 years
Solar WH	Can be used as primary or for preheating of water; solar exposure and orientation necessary for efficiency			High	Improved, dependent upon other factors; payback time can be as low as 2 years
Indirect WH	Coils run through a boiler or furnace (used to provide a building's heat) and then to a hot water storage tank			Moderate	Can be the most cost effective if a boiler or furnace is used for a good part of the year
Desuperheater	Draws heat from a source where heat is produced as a by-product, such as a bulk tank or air conditioner; water is preheated using this "waste" heat, saving costs by greatly increasing the efficiency of the primary water heating system				Very cost effective if heat source runs frequently
*Depending upon model's capacity rating.					

Information on Water Heating Systems

The U.S. Department of Energy (DOE) has an up-to-date website that discusses most of the pros and cons of typical residential and small-business water heating systems. Go to www.eere.energy.gov and follow the links. When studying the different options, keep the following questions in mind:

1. What is the maximum volume and maximum temperature required at any given time of use?
2. What is the initial cost versus the long-term operating cost?
3. Are there renewable energy rebates and incentives in my state that will help reduce the initial investment?

Table 7-1 breaks down the main features of most systems. The only one not included in the USDE's website is the desuperheater. Desuperheaters harvest "waste heat" from heat pumps, bulk tank compressors, etc., to help heat water. While not common, they are worth considering.

Reducing Use

Simple solutions apply when considering saving on hot water usage: Using water at the correct temperature for the need, insulating lines to maintain water temperature in the pipes, and reducing overall use whenever possible are all viable options for reducing hot water use.

Wastewater

Wastewater from your creamery can be divided into three categories: parlor wastewater, containing some animal waste as well as chemicals from cleaning; creamery wastewater, containing clear wash water and chemicals along with small amounts of milk, whey, and curd; and blackwater from the toilet/restroom. State and local regulating agencies may differ greatly in how they regulate each of these types of wastewater. You will hear about a great variety of methods that have been approved for dealing with them. Be sure to contact the correct authorities before proceeding with a wastewater management plan.

Calculating Volume

In most states you will need to estimate your wastewater volume as part of your sanitation application for septic, or for the animal waste management plan (AWMP) that some states require for even a very small dairy (see "CAFO and the Dairy Wastewater Management Permit" in chapter 2). You can make this calculation using the same formula as above for fresh water usage, since the same water will be running down your drains as waste.

Management Options

How you manage the wastewater from your creamery will depend upon the regulations of your local jurisdiction as well as your state. For the very small dairy,

Cheese curd being placed into forms for draining on Pholia Farm's draining table—made from a commercial dishwasher drain rack with stainless steel legs added.

many states defer to the local agency in regard to wastewater management. Smaller creameries will have very simplified wastewater management plans in comparison to the medium- to large-size dairy. Concentrated Animal Feeding Operation (CAFO) regulations were designed to properly deal with the wastewater and runoff from very large animal facilities where animals are confined, such as feedlots, confinement dairies, and hog or poultry operations. Some states have implemented the same permitting requirements for *all* commercial livestock operations, regardless of size. You will have to make sure that you are very clear on what your state and local authorities will expect and allow!

Below are some of the wastewater management options that *may* be allowed in your state. These options are by no means inclusive: Many other options exist, and exciting new ones are being developed, such as constructed wetlands

> ## MAKE IT GO A-WHEY!
>
> Whey, a natural by-product of cheesemaking, is considered an effluent if it is dumped down the drain, while in large facilities it can become a value-added product if quickly dried and sold as a food additive, protein powder, etc. When introduced to septic tanks, waterways, and wastewater, however, whey has a high biological (or biochemical) oxidation demand (BOD), making it a pollutant. Substances with a high BOD consume oxygen that is dissolved in the water. In plain terms, whey disposed of in water will create "dead zones"—not something you want in your septic system, stream, or pond.
>
> So what are your options? Most small creameries deal with the whey through two main disposal methods: feeding to animals and/or spreading on fields. If your whey is not too acidic ("sweet whey," with a pH above 6.0, is produced during the manufacture of cooked/pressed cheeses, such as Cheddar and Gouda), then it can successfully be fed back to your herd. Acid whey, with a pH around 4.6, is the by-product of lactic-fermented, soft cheeses, such as chèvre and ricotta; it can be fed to pigs and chickens but should not be fed to ruminants, as it could upset rumen acid balance. It can be diluted to alleviate some of these concerns and buffers can also be added, but this should be done with care. Some whey (sweet) can be used to create other cheese products, such as whey ricotta and Gjetost, but it is unlikely that you will be utilizing all of your whey for such products. Most creameries will find that the large volume of whey produced will require at least some of it to be disposed of, along with creamery wastewater, and spread on field, crops, or compost heaps.

(Living Machines built an indoor version at Cedar Grove Cheese in Wisconsin that processes 7,000 gallons of dairy wastewater per day), which send wastewater through a series of ponds or holding tanks where plants, microbes, and even fish cleanse the water naturally. For the purposes of this book, though, we will cover the more common systems.

Direct Diversion

This method is sometimes called "run-to-daylight." Wastewater basically goes down the drain and runs through a pipe to the ground's surface. This method, even where allowed, is not one I recommend, as it does not deal properly with the nutrients and chemicals present in the water; nor does it reuse the water as well as it could. Some filtration can occur if the water is run to a gravel and sand pit, but the longevity of this system cannot be predicted.

Septic System

Many jurisdictions allow for the use of a domestic-type septic system (either shared with a bathroom or independent). A grease trap should be included on the tank, and the drain field must be sized properly to deal with the large volume of water generated by the dairy. A properly functioning septic system allows for the disposal of solids (by tank pumping) and for the return of water to the ground.

Holding Tank

Holding tanks are the most common means of dealing with wastewater in newly constructed small dairies. A holding tank for wastewater is any type of

Wastewater holding tank, Fairview Farm, Oregon.

cistern, either buried or aboveground, that can receive and hold the wastewater. Periodically the water is distributed to crops and fields or disposed of in other acceptable fashions. A dairy's holding tank is basically a "dark" greywater system (as it will include small amounts of animal waste from the milking parlor) with a higher concentration of detergents and sanitizers than in a domestic greywater system. In addition, the distribution of the dairy wastewater will most certainly

FAIRVIEW FARM

Cheesemaker Laurie Carlson of Fairview Farm in Dallas, Oregon, describes her creamery's wastewater management system this way:

"We are newly licensed, so I'll describe our system. We had to put in an aboveground tank—1,550 gallons—to hold wastewater from milking during rainy months, so no water from the milking parlor would run onto wet pastures. We have a sump pump in the ground beside the milking parlor, which the floor drain runs into, then it pumps out to the tank. During dry periods we can run the water out of the tank onto the pasture through a hose. We are permitted with the condition that we do not milk during January through March, when our does are dry. If we planned to milk year-round we would have needed a bigger tank, because the winter months are the rainy ones.

"We had to get a CAFO permit here in Oregon, and we have a permit for 50 goats on the property.

"Our cheese processing building had to have a restroom with its own septic tank (commercial size, of course! For *one* toilet!). The floor drain in the cheese make room runs about 50 feet underground to a French drain in our orchard."

be regulated by CAFO rules. If you are interested in learning more about domestic greywater systems, go to www.greywater.com and www.thenaturalhome.com/greywater.html.

Ponds and Lagoons

Think of the dairy pond as an aboveground, open storage tank. Wastewater flows, or is pumped, to the pond, where it is held until it is spread. A lagoon is similar to a pond, but it's meant for longer-term storage and treatment of the wastewater. While most small dairies will not need the volume of storage space provided by a pond or lagoon, they are often seen on older dairies that have been refurbished. Water spread from a pond will have a higher nitrogen content than that from a lagoon (and consequently more odor). Both ponds and lagoons will require extensive waste management plans (through the CAFO permitting process) and engineering.

Power and Fuel

Too often, the energy needs of the small dairy are not adequately calculated before designing and outfitting the creamery. By not factoring in these costs early in the planning stages, the small business can find itself facing higher power bills than expected and unable to keep costs in line with income. With energy costs being variable, and with long-term forecasts of price increases, it is important to attempt to address energy usage needs not only before opening for business, but before adding any new piece of equipment.

> **ESTRELLA FAMILY CREAMERY**
>
> Kelli and Anthony Estrella of Estrella Family Creamery, Washington, saw their electric power bill go from $50 a month to $500 a month when their creamery came online. The family-run farm produces award-winning raw-milk and cave-aged cheeses from their herd of 19 Normandie cross cows and 50 La Mancha goats that graze the 164-acre restored dairy farm.

Calculating Usage

Don't be intimidated by such terms as watts, kilowatts, amps, and volts. You don't have to completely understand electricity to figure out how its use will affect your business. There is a simple formula that can be used to determine the potential energy usage of any appliance:

Formula to Determine Energy Usage
Volts × Amps = Watts

So if you don't really need to know what all of this means, how do you make it work for you? If you take a look at any electric appliance, you will find a plate or label that states the volts and amps (short for amperes) that the unit uses at peak usage. When you multiply these two numbers together, you will find out how many watts the equipment will use at any given time. To take it one step further, if the equipment runs at that rate for 1 hour, its usage could be measured in kilowatt-hours (KwH). Your electric bill tells you how many kilowatt-hours you use per month and how much those hours cost you. One kilowatt-hour is equal to 1,000 watts. For example, if a 100-watt light bulb runs for 1 hour, it will use 0.1 KwH of electricity. If it runs for 10 hours it will use 1 KwH.

> **PHOLIA FARM**
>
> Before we built our off-grid dairy, creamery, and home, we plugged every appliance we had into a Kill A Watt meter to find out how much power we would have to make. Our example is extreme, but it is an eye-opening experience to find out how much power some things really use!

Here is an example of using the formula of *volts × amps = watts* for our milking machine:

- 115 volts × 6.6 amps = 759 watts
- 759 watts × 2 hours of use = 1,518 watts or 1.5 KwH
- 1.5 KwH/day = 45 KwH/month
- 45 KwH × [cost per KwH] = electricity cost per month
- If the average price per KwH across the U.S. is $0.11, then it will cost you about $5.00 per month to operate this equipment.

Here is another example, using a more power-thirsty appliance—a three-compartment, glass-fronted refrigerator:

- 115 volts × 12 amps = 1,380 watts
- 1,380 watts × 10 hours (as the compressor will not run all of the time) = 13.8 KwH
- 13.8 KwH/day = 414 KwH/month
- 414 Kwh × $0.11/hour = $45.54/month

You can see how this usage might all start adding up to a sharp increase in your power bill!

A very useful tool to get an even more accurate idea of how much electricity a piece of equipment is really using is a neat gadget called an energy usage meter. A popular brand is called the Kill A Watt. You plug the meter into any 110-volt wall socket and then plug an appliance into the unit. A digital readout will display the current usage, as well as usage over time. Wait about 24 hours and you can get a pretty accurate idea of how much power is being consumed—and think about whether adding that second fridge is really worth it!

Reducing Needs

When you design your creamery (as well as your business plan), you can make power-saving choices that will positively impact your future power bills and environmental footprint. The following chapters will discuss options that you can build into your plan for energy efficiency. This can be a greater challenge when you are working with existing buildings and infrastructure, but you can still make choices that will help reduce your energy needs.

If you are considering increasing production in order to increase income, don't forget to factor in increased energy usage and costs. If you are already using your equipment at its maximum and you need to add additional cold storage, etc., be sure that the increased product will offset the increased costs enough to make the investment worthwhile.

Renewable Energy

Renewable energy (RE) can be expensive to purchase and install, but it can become a cost-saving investment for any size creamery. The USDA Farm Bill often includes grant moneys for farmers who want to use RE. Check its website for updates and links to state officials who can help you research these options: www.rurdev.usda.gov/rbs/farmbill.

Another useful site for monitoring state incentives as well as other funding possibilities is www.dsireusa.org.

Tax credits and state rebates can sometimes cover as much as 50 percent of your initial RE purchase and installation costs, but these vary from year to year and state to state. Finding a reliable RE installer will be your best bet for staying on top of state funding and how to best take advantage of it. Many RE companies now offer financing, as well. Think of it as buying your power in advance—for a set price now, you are guaranteeing a steady source of power for years to come.

Often people focus on the "payback time" for renewable energy (the time it would take, with conventional operational costs, to pay for an RE system). Payback time isn't usually a consideration when using conventional energy sources, buying groceries, or buying a car—so if you're considering renewable energy, take the

GIANACLIS'S "CUPS" FOR MAXIMIZING ENERGY EFFICIENCY IN THE CREAMERY

1. **Choose the right-size equipment.** For example, don't buy a 100-gallon bulk tank if you will be storing only 50 gallons of milk.
2. **Use equipment to its maximum capacity for its size.** In other words, don't run a large refrigerator only half full of product.
3. **Place equipment where it can run efficiently.** For instance, don't place a freezer against a wall that receives full sun exposure and radiates heat. Place cooling units, such as bulk tanks and refrigerators, so that the excess heat they create (as a by-product of cooling) can be exhausted from the room (to increase the efficiency of the compressor).
4. **Schedule processing to take advantage of natural conditions.** For example, if the make room is so warm in the afternoon that you have to run an air conditioner, make cheese earlier in the day.

A micro-hydroelectric turbine provides additional power to Pholia Farm's off-grid solar power system.

payback time information with a grain of salt and choose what works not only for your pocketbook, but for your philosophy.

For a detailed description of renewable energy options for farms, go to the National Renewable Energy Laboratory's website and visit the page for farmers and ranchers at www.nrel.gov/learning/farmers_ranchers.html. Another great site for farmers considering RE is www.farmenergy.org. While some of the RE options listed, such as methane recovery systems, are currently aimed at the large-scale dairy, others—including wind and solar (photovoltaics)—are viable options for any size farm. Determining your site's best choice and system size is most likely a job for a licensed installer.

Structure

As you think about building a new barn and creamery, you will have options for choosing designs and materials that can increase your buildings' efficiency and reduce maintenance. While traditional (conventional) materials, such as wood framing and cement block, are always options, these days it's becoming easier to find contractors who have experience with some interesting, often more ecologically friendly options—such as structural insulated panels (SIPs) or insulating concrete forms (ICFs). New construction will also give you the opportunity to orient the building and position windows to maximize efficiency. Even if you are remodeling, you might be able to take advantage of some of these options.

Building Material Choices

There are many options available for the construction of a small dairy. Some of these choices may not be available in your area, either due to lack of supplies and construction expertise, or due to building codes that have not evolved to accept some of the newer building choices.

When you choose a building material, it is a good idea to consider the environmental cost (EC). Environmental cost figures attempt to factor in such things as carbon output during production of materials, man-hours of installation, life span of the material, and demolition and disposal costs. Bear in mind that data on these factors is subjective and constantly changing. Depending upon where you live, the intended use of your building, the expected life span of the business, etc., the actual EC will vary greatly.

Here is an example: Concrete has a high up-front environmental cost. One of the key ingredients in concrete is cement, which is made from limestone that is mined and then kiln treated. Much energy is consumed not only during the process of producing concrete but also during its use as a building material—including trucking and often pumping the concrete to, and at, the building site. On the plus side, however, since any building will use far more energy over its life span than during its construction, concrete's long-term assets—which include thermal mass (the ability of a material to store heat) and low maintenance (cleaning, upkeep, paint, etc.)—add EC credits. Furthermore, at the end of its life span concrete can be recycled for new uses. In addition, its fire-resistant quality adds sustainable benefits (as well as insurance rate discounts) that can all add up and make it the greenest choice available—in the appropriate application. (If it sounds

TABLE 7-2: Comparison of Common Building Material Options

Material	Cost Comparison	Pros	Cons
Stick (Wood Framed)	Least	Accepted by all building codes; common usage means more experienced builders	Lower energy efficiency, increasing cost of materials; fire and termite concerns
Steel Frame	Moderate	Accepted by all building codes; common usage means more experienced builders	Higher initial cost than wood
Cement Block = Concrete Masonry Unit (CMU)	Moderate	Durability, low maintenance	Low R-value*
Insulating Concrete Forms (ICFs)	Moderate–High	High R-value*, durable wall surface	Higher electrical and plumbing installation costs
Structural Insulated Panels (SIPs)	High	High R-value*, minimizes use of wood and waste at jobsite, fewer man-hours**	Does not allow for changes in floor plan during construction; higher plumbing and electrical installation costs

*R-value is a rating used to indicate a material's ability to prevent the transference of heat and cold. Higher R-values mean more insulation.
**Fewer man-hours does not necessarily translate into a cost savings.

a bit confusing, it is, but it is still a good idea to learn what you can and make the best choice possible—when options exist.)

Table 7-2 summarizes some of the primary differences in building materials.

Passive Solar Design

Passive solar design optimizes a building's design to suit its climate and location to help maintain optimal comfortable temperatures year-round. Factors to be considered include the orientation of the building (south-facing usually being ideal); thermal mass of the materials (walls and surfaces that are placed to intentionally retain seasonally desirable temperatures); windows placed and sized to minimize undesirable heat loss or gain; and shading, such as deciduous trees and awnings that keep the building cool in the summer and allow sun exposure for increased warmth in the winter. Existing buildings, geographical limitations, or other factors can limit the use of passive solar design, but if you keep it in mind and incorporate it whenever possible, you will reap a great benefit on your power bill!

Ergonomics and Efficiency of Motion

At this point, you are probably feeling as if there is enough to think about when designing your creamery, but here is one more! Think of sustainability in terms of how you might design your workspace to maximize your efficiency and therefore, in the long run, the sustainability of both your business and your enjoyment of the work. In the beginning, we all just want to do whatever it takes to get our license and start selling cheese, but some of the choices we might make to arrive at this short-term goal as quickly as possible could adversely affect our long-term survival. So, whenever possible, make choices that will reduce your workload and increase your efficiency! This topic will be revisited as you continue reading about designing each room of your creamery.

As you continue reading part 3, I hope you will see how your floor plan design and equipment choices will affect your choices of infrastructure and system options. No one I know who has built a creamery has been able to look back and say, "It's just right; I wouldn't change a thing." But with some thought and consideration, perhaps your creamery will be close to ideal.

· 8 ·

The Milking Parlor and Milkhouse

What transforms a farm into a dairy? Two rooms: the milking parlor and the milkhouse. Both of these specialized rooms must be constructed to meet regulatory standards as well as to suit the size of your herd and your own personal preferences.

For both rooms we will cover floor plan considerations, construction and maintenance standards, and equipment options. Appendix B contains sample floor

Eager to be milked, these pasture-fed cows wait for their turn at Sweet Home Farm, Alabama.

plans from working farmstead creameries around the country—feel free to refer to them during your reading. Appendix A provides a list of resources where you can get much of the equipment mentioned here.

The Milking Parlor

While the term "parlor" may bring to mind fringed lamp shades and overstuffed furniture, the milking parlor is in reality a bustling, noisy, wet room that should be designed to withstand the abrasion of animal hooves and the harsh contact of cleaning chemicals and brooms.

Floor Plan Considerations

When designing the layout of your milking parlor, take into consideration access, location, and size.

Access

Milking takes a huge chunk of time out of the dairy farmer's day. Getting animals in and out of the parlor in an efficient fashion will cut down on wasted time and effort. Think about animal traffic flow when designing the floor plan of your milking parlor, especially in relationship to the loafing area, holding pen, and barn. If you will be milking by hand or using a portable milking machine, then easy access to the milkhouse is a priority—lugging heavy pails of milk up steps and through doors will get old very quickly!

Location

Locating the parlor close to or adjoining the barn brings convenience, but also the added challenge of keeping barn material—such as manure, flies, hair, and dust—out. If milk will be transported to a bulk tank via a permanent pumping system, locating the parlor slightly above the bulk tank will allow for gravity-assisted pumping—and the milk will be treated more gently.

> **TRAFFIC FLOW**
>
> Paul Hamby, owner of Hamby Dairy Supply, one of the nation's largest suppliers to home and commercial dairies, says: "Traffic flow is perhaps the single biggest mistake people make in planning a new facility. Poor flow adds more labor and slows the milking process and usually cannot be fixed after your facility is up and running."

Size

As with most rooms in the dairy, it is hard to make them too large. A good parlor should accommodate the animals, the milkers, and the occasional additional person, such as a milk tester, visitor, apprentice, etc.

Construction and Maintenance Standards

These standards are defined in the Pasteurized Milk Ordinance (PMO), the FDA's bible on Grade A milk production. Each state may also dictate additional and

> ### WILLOW HILL FARM
>
> Willow Smart and David Phinney of Willow Hill Farm in Milton, Vermont, built their new creamery on a slope—with each room stepped down a level from the one preceding it in the cheesemaking process, thus allowing them to utilize gravity at each stage. Even the whey flows out of the creamery through a pipeline that runs directly to the hog pens. Willow Hill milks a herd of approximately 90 East Friesian and Friesian cross ewes and 7 Brown Swiss and Dutch Belted cows, which provides them with enough milk to produce 15,000 pounds of cheese per year. Their new creamery includes a handy self-serve "store" from which customers can also view the cheesemaking room.

unique standards, so be sure to consult with your state's regulatory agency prior to building. This book will refer to the standards in the PMO at the time of writing.

Floors
- Sloped to drain
- Made of concrete or other impervious material

The parlor floor is not expected to be as smooth and free of blemishes as other rooms in your dairy, but it should not have areas that allow water to accumulate in puddles or cracks that are too deep to clean. You can have a slightly rough texture to your parlor floor to prevent animals or milkers from slipping. Parlor floors, especially those used by cows, will need to be resurfaced periodically over time (usually many years).

Walls and Ceilings
- Cleanable surface
- Sealed junctions at floor and ceiling

> ### SWEET HOME FARM
>
> One of the more clever ceiling materials I saw was at Doug Wolbert and Alyce Birchenough's Sweet Home Farm in Alabama. Doug used painted metal roofing to cover the ceiling interior. He sealed screw heads and joints with 100 percent silicone caulking. The material has held up extremely well in the moist environment and looks clean and fresh with minimal maintenance. Doug and Alyce produce approximately 13,000 pounds of cheese annually from their herd of grass-fed Guernsey cross cows.

Walls and ceilings must have a surface finish that you can keep reasonably clean. Concrete and block will have to be filled and relatively smooth. Ceilings should be smooth and painted or finished in an approved manner. Ceiling-to-wall joints as well as light fixtures must be sealed to prevent dust from falling into milk (especially when feed or other things are stored above the parlor). You should expect to resurface concrete floors periodically, depending upon what kind of wear and tear they experience. (For example, concrete floors in a cow parlor will wear much more quickly than those in a goat parlor.)

> ## GRADE B MILK
>
> While the PMO covers the production of Grade A fluid milk, some states still license plants to produce Grade B milk, also known as "manufacturing milk" because of its use in the manufacture of other dairy products, including cheese. Regulations differ for Grade B in regard to dairy barn construction and bacterial counts allowed in milk. Inspections are not as frequent. Some farmers start out using Grade B to make cheese, with the thought of upgrading in the future. Before you start your construction, consult with your licensing agency to determine if Grade B is an option worth investigating in your state.

Lighting

Lighting must be sufficient for well-illuminated day or night milking. The PMO stipulates a light meter reading of 110 lux. (Think of a living room with enough light to read easily.) Light fixtures should be sealed to prevent any broken bulbs from introducing hazards into the milk. The most acceptable light fixture consists of a bulb inside a rugged glass jar encased by a metal grid. These fixtures are somewhat costly, but they'll reduce the risk of hazards due to broken bulbs (which can occur easily when cleaning the ceiling and walls with a scrub broom).

Ventilation

Good ventilation is important to prevent condensation or excessive odors. Consider installing an exhaust fan with a timer if natural ventilation through windows is not an option year-round.

Openings

The PMO allows for three-sided parlors (ones that are open to a holding pen, breezeway, or other livestock access), but remember to check the rules for your specific state to see if this is acceptable where you are. Any opening to a feed room must have dust-tight doors that are closed except when they are in use, to prevent the contamination of feedstuffs into the milk. In addition, poultry, swine, and pets may not have access to the parlor. For large openings, using vinyl strip curtains or doors can help with climate and pest control.

Equipment and Accessories

Here is a list of some of the equipment and extra stuff you will need to consider for your milking parlor:

- Milk stand or platform
- Milking machine system, portable or in-place
- Hand-washing sink and wash-down hose
- Other stuff: shelf for teat dip, udder wipes, radio, white board, calendar, clock, etc.

Milk Stand or Platform Considerations
- Number of animals that need to be milked—both now and planned over the next five years
- Construction material, cost, and maintenance
- Parlor style: tie-stall; pit/recessed (with herringbone or parallel pattern); side exit; or rotary/carousel

Number of Animals
While it is often not possible to anticipate future growth, try to start out with a milking stand or platform that will anticipate your needs about five years down the road. This could save significant cost as well as time should your herd size grow. Also, remember that the more animals you can milk at one time, the faster your milking chores will be completed!

Construction Materials
If you are building a recessed (pit) parlor, then you will likely build your platform from concrete, most certainly if you are milking cows. If you are milking goats or sheep, you can have a recessed parlor with the elevated platform fabricated from steel—either painted, galvanized, or stainless steel. (If you are milking year-round, it is a good idea to spend the extra money on a hot-dipped galvanized surface or even stainless steel.) A handful of states allow the use of painted wood. Some have argued that painted wood is an impervious material and should be allowed, but in reality, a painted surface on which hooved animals will be walking will be in constant need of maintenance.

Parlor Style
In the Northeast and colder parts of the nation, you will find many *tie-stall* barns that both house the cows during the winter and serve as a milking parlor. These are rarely seen in more temperate parts of the nation, where animals can access an outdoor loafing shed or protected paddock year-round. In some parlors, animals are not given grain during milking and can be clip-tied to either a wall or a rail. While this does eliminate the cost of purchasing a headgate system, it also increases labor time, as each animal must be individually secured and released.

Recessed or pit parlors can be designed with either a herringbone or parallel formation:

- A *herringbone* parlor usually consists of a central pit flanked by two milking platforms. Animals enter on one side and line up diagonally. The milker prepares the first set of animals and begins their milking, then allows a second set into the other side and gets them started. After being milked, animals exit from each side at the same time.
- In a *parallel* parlor, animals line up side by side at a right angle to the milker. Some parallel systems allow for single exit of animals, while others allow group exit.

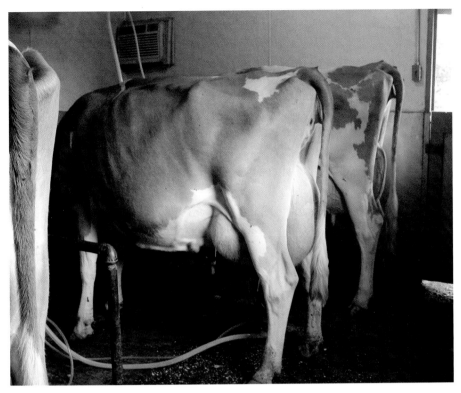
Three-cow tie-stall parlor, Sweet Home Farm, Alabama.

Six-sheep milking parlor with cascading headgate, Black Sheep Creamery, Washington.

> **HEADGATES**
>
> For the small goat or sheep producer, a parallel formation on an elevated platform is the most common parlor design in use. The does or ewes usually place their heads in a stanchion (headgate) and are secured during the milking process. Headgates can be gang-operated (a single mechanism locks all animals in at one time); cascading (the first animal can access only the first stanchion; as it places its head through the opening, the stanchion locks the animal in and triggers the next stanchion in sequence, to open, and so on); or individually activated (each station closes and releases independently of the rest). Cascading headgates have, in my opinion, the advantage in that they prevent animals from stopping at the first open stanchion and blocking animals behind them from entering the parlor. The only catch is that in order for the cascading headgate to latch, the animal must lower her head to a grain tub, thus triggering the catch. So you must want to feed the animals grain while they're on the milk stand for this type of headgate to be a good choice for your farm.

Side-exit parlors have individually loading and exiting stalls that place the cow sideways to the milker.

Rotary or carousel parlors are just what they sound like: animals step onto a slowly rotating platform and face toward the center, placing the udder to the outside.

As you can imagine, there are many things to consider when choosing a parlor style. There are advantages and disadvantages to each type. Some of the main points to consider when choosing a parlor style are:

- *Ease of access to the animal for udder prep and milking.* The easier the access, the more sanitary the process is likely to be.
- *Individual attention.* If animals must come and go in groups only, then slow-milking animals will either impede the process or will not be completely milked out when they exit.
- *Visibility for the milker.* The farther the milker is from animals during milking (such as in a larger side-exit parlor), the more likely teat cups will not be removed at the right time.
- *Comfort of the milker.* Parlors designed for one species (such as cows) but used for another (such as goats) are likely to be less efficient and comfortable. Platform height that is too low for the milker to easily prep and milk animals will decrease comfort and therefore efficiency.

If possible, try to visit dairies using several of these systems to get a better sense of your options. Consulting a specialist who sells parlor systems can also be helpful; however, many of these companies are focused on systems that are designed for the large dairy.

Milking System Considerations

While the PMO does allow for hand-milking, there are very few dairies that still employ this method as their primary milking system. Most small operations

A "Semi-Permanent" Bucket Milker

A bucket milker can be set up to be "semi-permanent" by locating the vacuum pump and motor a distance from the milking parlor and installing a PVC line that runs from the vacuum pump to the milking parlor and an electrical line and switch for turning the motor on and off in the parlor. This allows for a quieter milking situation, and easier cleanup in the parlor; it also keeps any greasy residue from an oil vacuum pump out of the milking area.

Rhonda Gothberg setting up her semi-permanent bucket milking system at Gothberg Farm, Washington.

utilize either portable bucket milkers or a permanent pipeline system. Important factors to consider when choosing a milking machine system are:

- Number of animals to be milked, now and over the next five years
- Initial cost and setup
- Operator efficiency and comfort

Portable Bucket Systems

Portable bucket systems consist of a vacuum pump, a pulsator, a milk bucket, milking lines, and a milking cluster made up of claws and teat cups. The vacuum pump has a line that runs to the milk bucket, where it creates negative pressure. The milking lines run from the bucket to the animal. A *pulsator* is installed either on the lid of the bucket, in the vacuum line, or as an all-in-one claw and pulsator. The pulsator turns the vacuum suction to the milk bucket and teat cups on and off rhythmically, allowing the teat to refill with milk and ensuring that circulation of blood to and from the teat is not impaired. The vacuum pulls the teat into the teat cup, which consists of an *inflation* and *shell*. The inflation has a silicone or rubber liner inside a plastic or stainless steel shell. At the bottom of each inflation, or in-line just after the inflation, is a part referred to as the *claw*. If you are milking cows, the claw looks a bit like an animal claw, with many attachments coming off a central part, but for goats and sheep there is no resemblance. The whole setup is referred to as the *cluster*. The claw consists of a reservoir where milk pools before it enters the milk hose, a valve that prevents backflow of milk into the inflation, and a tiny air vent to draw air into the line—thus preventing "slugs" of milk in the lines (slugs are a problem because they cause the vacuum to fluctuate to an extreme that can allow for droplets of milk to spray upward and into an open teat orifice). If no claw is used, then venting must be provided through the shell. Both the milk line and a vacuum line from the pulsator attach to the claw.

After milking, the bucket is transported to the milkhouse, where the milk is poured through a filter (unless an inline filter is used) into a clean milk can or bulk tank; after this, the milk is either chilled and stored or pumped directly to the cheese make room for processing. Portable systems are cleaned using hose brushes and a sink and/or bucket washer (more on this in the section on the milkhouse).

Pipeline Systems

Pipeline systems deliver the milk directly to a milk storage (bulk) tank, eliminating the need to transport buckets by hand. The main components (milk pipeline and vacuum lines) are built in, while milk hoses and clusters can be either cleaned-in-place (CIP) or cleaned separately in the milkhouse. Most newer pipeline systems are CIP, but many smaller farms or dairies that milk in tie-stall barns remove milk hoses from the pipeline and clean them manually or with a bucket washer in the milkhouse. Pipeline systems use the same cluster set-up as described for portable bucket milkers, but instead of the milk hoses being

TABLE 8-1: Milking Machine Cleaning and Maintenance

Part	Life Span	Daily Cleaning	Periodic Cleaning/Maintenance
Pulsator	Rebuild (replace rubber parts and gaskets) when needed; depends on cleaning and hours (can be 6 months to several years); keep a rebuild kit on hand!	Rinse, brush with detergent, inspect intake for signs of milk entering pulsator	Weekly disassembly from bucket lid or line to clean and inspect
Plastic Milk Hoses	4 months to 1 year	Lukewarm rinse, alkaline detergent wash, acid rinse	Disassemble, clean with hose line brush as needed
Silicone Milk Hoses	1–3 years	Same as plastic	Same as plastic
Rubber Inflations	3 months	Same as milk hoses	Same as milk hoses; use inflation brush
Silicone Inflations	1 year	Same as milk hoses	Same as milk hoses; use inflation brush
Claws	Several years; inspect for cracks; replace internal gaskets annually	Same as milk hoses; inspect air intake vent and clear of any blockage if needed	Same as milk hoses; annually replace rubber gaskets if present

attached to a bucket, they are attached to a milk pipeline and the vacuum lines are attached to a main vacuum line. These main lines run either above the level of the udder (high line) or below (low line). It is generally agreed that low-line systems are more efficient and move the milk more gently, as they can rely more on gravity to aid in the milk flow. In the small dairy, teat cups will most likely be removed by hand when the animal has been emptied of milk, but in large systems, automatic take-offs (ATOs) are common. The milk flows through the pipeline to a milk receiver jar, where it is pooled and then pumped or released to the bulk tank for storage. There are many variations, modifications, and maintenance issues that exist with a pipeline system. (These include such things

BACKFLUSHING AND MASTITIS PREVENTION

In large dairies, pipeline systems are more complex and automated than you are likely to see in a small dairy. This includes automatic take-off (ATO) and automatic backflushing of the teat cups. Backflushing is the process in which a sanitizing solution is flushed through the teat cups to remove any milk residue from the animal just milked and sanitize the teat cups between animals. Since the solution flows in the opposite direction of the milk, it is called backflushing. The removal of the milk residue can help prevent cross-contamination by mastitic agents. Manual backflushing can be done on any milking system by shutting off all vacuum to the cluster and immersing the teat cups in a sanitizing solution; however, exposure to the solution must be for the minimum time required (30 seconds for most properly diluted sanitizers), and great care must be taken to refresh the solution (as it will rapidly become ineffective as the milk residue is rinsed into it) and to ensure that no solution is pulled into the milk hoses. Most references to manual backflushing as a means of preventing the spread of mastitis suggest that other steps will be equally effective, including ensuring proper vacuum pressure; not allowing teat cups to slip on and off a teat during milking (causing milk in teat cups to possibly enter the teat orifice); proper removal of teat cups, with vacuum off; and proper teat sanitation.

Eight-doe pipeline CIP milking system, Fraga Farm, Oregon.

> ### MILK QUALITY OPPORTUNITY
>
> Your best defense against poor milk quality is aggressively enforced sanitation and proper procedure in the milking parlor. Not only will these good manufacturing procedures (GMPs) and Standard Sanitation Operating Procedures (SSOPs) ensure a clean milk supply, but they will help prevent cross-contamination between animals and the spread of a potentially herd-devastating disease. The correct operation of vacuum pumps and milking equipment will help maintain good udder health, without which quality milk cannot be produced! See appendix C for more on milk quality and chapter 12 for more on SSOPs and GMPs.

as automatic backflushing of teat cups, in-line cooling and filtration, and more.) You would be well advised to consult with a reputable dairy supply dealer when considering the purchase of a pipeline milking system and then maintaining it.

Choosing the Right System

The more animals you are milking, the greater your volume of milk—and possibly the longer it will take to relieve them of this milk. This brings into consideration two important factors: First, the greater the volume, the more difficult it becomes to transport the milk by hand. Second, the longer it takes to complete the milking, the less quickly milk will be chilled (if not automatically transported to a chilling tank). A portable bucket system is very inexpensive in comparison to a pipeline system, but there could be added labor costs if the volume of milk reduces operator efficiency.

The two system types are virtual mirror images in their advantages and disadvantages. Here is a little quiz that might help you choose:

1. How many animals will you be milking?
    a. 1–20 goats/sheep or 1–3 cows? *Portable okay.*
    b. 20–30 goats/sheep or 4–6 cows? *Portable okay, but pipeline better.*
    c. 30 or more goats/sheep or more than 6 cows? *Pipeline best.*
2. Is your milking parlor already built?
    a. Yes? *Consider portable.*
    b. No? *Consider pipeline if milking more than number of animals in (1a) above.*
3. How much money can you invest?
    a. Minimal? *Choose a portable system.*
    b. A bit more than "not a lot"? *Consider a pipeline for long-term benefits if milking more than 20 goats/sheep or 3 cows.*

If you decide that a portable system is best, there are a few things to keep in mind when choosing a system: motor size, oil or oil-less vacuum pump, balance tank, and maintenance issues. Motor size, measured in horsepower (HP), will dictate the maximum number of animals (or clusters) that can be milked at one time. If the motor doesn't have enough HP, then it cannot create sufficient vacuum without working too hard and shortening its life. With respect to oil lubrication of the vacuum pump, you will find equal support for either being "the best," but remember that an oil pump will spew out an oily residue that will need to be factored in for placement of the unit. Balance tanks serve several purposes on a milking system. First, they provide a cushioning effect on the vacuum supply, preventing or limiting severe fluctuations in vacuum when lines are taken on and off of animals. Second, it provides a distribution chamber for line from the pump and to the milking clusters. Last, it serves as a trap (or leads to a trap) for any milk that might inadvertently get sucked into the vacuum lines. Some very small portable systems made today do not have a balance tank. If any milk enters the system, it can go directly into the vacuum pump and cause damage. With any portable, or, for that matter, pipeline system you will need to know the proper maintenance schedule. Be sure to consult with the dealer and know what is needed to keep your system operating at its peak.

As with parlor design, it is helpful to have some hands-on experience with milking equipment options before you choose and invest in a system of your own. Discuss the choices with people who have used both, as well as with vendors who sell both. While you can always upgrade from a portable to a CIP system, it may save you time and effort to install one in the beginning if need will eventually demand it.

Hand-Washing Sink and Wash Hose

A hand-washing sink is not required in the milking parlor, but consider installing one anyway, as it does come in handy. Parlor floors must be kept clean. While a washdown with water is the most common method, dry brushing followed by the application of hydrated lime (to sanitize the floor) is allowed by the PMO. Goat and sheep parlors can easily be swept first, to minimize the amount of solid waste entering the wastewater management system. The wash hose does not need to be supplied with hot water; however, you might want to consider having hot water accessible for more intense cleaning.

Other Stuff for Your Milking Parlor
You can outfit your milking parlor with other things that will increase your comfort and functionality. A nearby shelf or a rolling cart (get the plastic/rubber type) for such items as teat dip, wipes, and udder treatments is very useful. A radio or music system is nice for both the milker and the animals. A calendar with important breeding and treatment dates, along with a whiteboard to leave messages for the next milker (or for yourself), can also help optimize the time spent in the parlor by increasing the efficiency of your herd management.

The Milkhouse

The room that houses the bulk tank, and therefore the milk, is called the milkhouse. While a small creamery may actually store its milk in the cheese make room, it will still be required to have a milkhouse for the cleaning and storing of milking equipment.

Floor Plan Considerations
When designing the layout of your milkhouse, take the following points into consideration:

- *Access* that makes it easy to bring milk and milking equipment in from the parlor; when allowed, have doors swing into the room for easier load transportation.
- *Location* that is convenient to the parlor and make room, but separated for reasons of cleanliness and sanitation.
- *Size* that allows for ease of work and cleaning, along with possible equipment upsizing over time.

Construction and Maintenance Standards
As with the milking parlor, the construction and maintenance standards in this book conform to the PMO at the time of writing. Remember that your state may enforce different requirements; therefore, be sure to consult with your regulatory agency prior to any construction.

Floors
Floors must be sloped to drain and made of concrete or other impervious material. They should be finished smooth and free of cracks or blemishes that inhibit

proper cleaning. Floor drains must be accessible for cleaning and inspection. Don't put a drain where equipment will sit!

Walls, Ceilings, and Doors
Wall finishes in the milkhouse will need to be more easily cleaned than those in the milking parlor; in other words, they will need to be smoother and in better repair than might be acceptable in the parlor. Painted plywood, fiberglass-reinforced panels, dairy board (also known as high-density polyethylene boards HDPE), and sealed and painted concrete (block or other) are all acceptable. Keep in mind that any paint going on a block or concrete wall will need to be formulated for wet environments. Many cheesemakers have had to completely strip and redo their walls after paint failure! Doors should be tight fitting and self-closing. If your regulators allow, try to design doors to swing inward to facilitate carrying loads into the room. Some state regulations will not allow for the parlor-to-milkhouse door to swing into the milkhouse, the concern being that an animal could push the door in and gain access to the milkhouse.

> **WHAT KIND OF DOORS?**
>
> This question seems so simple to me now, but during our construction it was a tough one. For some reason I thought we had to buy heavy-duty, commercial-grade doors (very expensive). Fortunately, we figured out that it didn't matter, as long as they were sturdy and easy to clean. We chose fiberglass-wrapped doors with reinforced interior panels. (Most fiberglass and steel doors have a foam insulation core.) We also had windows put in each one, allowing extra light as well as visibility when coming and going into the room.

Lighting
The milkhouse should be more brightly lit than the parlor. A lux of 220 is required. (Think kitchen lighting brightness.) Light fixtures should be protected and covered to prevent any broken bulbs from introducing hazards into the milk.

Remember, moisture allows for bacterial growth!
A dry environment is more likely to be a sanitary environment.

Ventilation
As in the parlor, there must be enough air volume and circulation to prevent condensation and allow tools and equipment to dry—remember, moisture allows for bacterial growth! A dry environment is more likely to be a sanitary environment. A good-quality exhaust fan (suitable for damp environments) with a timer switch (so it can run for a period of time after cleaning is completed) is highly recommended. Be sure to include a screened opening for fresh air to enter when the exhaust fan is running. Windows are permitted by the PMO (as long as they do not open directly to animal housing) and can supply adequate ventilation during most seasons.

Water

Hot water must be supplied to the milkhouse at a minimum temperature of 140–160° F. This is to allow for the proper cleaning of milking equipment. The water must also be of a sufficient volume and have adequate pressure for cleaning. Insufficient hot water supply is one of the most common mistakes made when designing systems for the dairy. While hot water may arrive into the milkhouse at an adequate temperature, it cools rapidly during pumping through milking lines and equipment. If water is not hot enough, solids will be redeposited during the wash cycle—this can lead to buildups of fats, proteins, and other contaminants that will harbor and grow bacteria, which in turn lead to the contamination of milk.

> **Insufficient hot water supply is one of the most common mistakes made when designing systems for the dairy.**

In addition to water temperature, water for cleaning should be of a close to neutral pH and not be too high in minerals, such as iron. While your regulatory agency will most likely test the water for coliform contamination, I recommend a test for minerals (especially calcium and magnesium, the main components of hard water) and pH. You can test pH with pH strips or a pH meter (the same can be used for cheesemaking). If the mineral content of your water is high enough to negatively affect cleaning and sanitizing chemicals, it could also be affecting the absorption of and interaction with other minerals your livestock need. Consult the company that supplies and/or manufactures your cleaning and sanitizing chemicals for parameters and product suggestions. Water quality and its effect on cleaning quality will influence the life span of your milk hoses and equipment, so it should not be ignored!

Equipment and Accessories

Here is a list of some of the equipment you will need for the milkhouse, followed by details on each item. The bulk tank can be located in the cheese make room, but we will cover it under this section.

TIP

Shelving

Wall shelving in the dairy, especially the milkhouse and creamery, can be challenging. Plastic-coated wire holds up for several years, but the brackets and mounts designed to hold it are made of painted steel. Any painted metal in the dairy will begin to flake and rust in a very short period of time. These flakes of rust and paint can become a contaminant in your milk and product. Chrome-coated shelves will also rust. Stainless steel won't rust in normal circumstances, but it will in the presence of high concentrations of chlorine. Epoxy-coated shelving is, in my opinion, the best choice. Even then, watch out for the mounting hardware (screws, etc.), as they will often rust out, leaving your shelves in a precarious position. Choose stainless steel screws and bolts whenever possible.

- Double-compartment sink
- Hand-washing sink and wash hose
- Milk cooling/bulk tank
- Milk line washing equipment
- Milk strainer and receiving pails (when using portable milking system)

Double-Compartment Sink

A two-compartment sink is needed for the washing and rinsing of milking equipment and tools. For the very small producer, the second compartment can be a stainless steel pail or a small, side-hanging sink as used with a bucket washer (more on that in a bit). You can hang your clean buckets and pails above the sink on stainless steel "S" hooks hanging from mesh shelving.

Hand-Washing Sink and Wash Hose

A hand-washing sink should be equipped with single-use towels (either disposable paper towels or cloth towels that are laundered between use). Be sure the sink stays clear of items that would block usage; in other words, don't leave baby bottles, pails, or anything sitting in it. Part of the routine dairy inspection will verify the access (and appropriate use) of hand-washing sinks. A wash hose with hot and cold water will be very useful, especially when cleaning tanks and hard-to-reach corners in the milkhouse. Don't forget to use a floor squeegee to reduce water waste (instead of using the water sprayer as a broom).

Milk Cooling (Bulk) Tank

There are cheesemakers who do not use any milk storage system; rather, they make cheese daily. While this is the best choice for producing high-quality cheese, it is not realistic for most farms that rely on only a few people to do all of the labor. If you make cheese daily, other important things, such as animal care and maintenance (not to mention your own personal life), could take a backseat. I will assume that most people reading this book will need to hold their milk for 24 to 72 hours on a regular basis.

For the larger producer, there are many bulk tanks available, both new and used. For the small dairy, however, the choices are very limited. New tanks currently manufactured in the U.S. start at 600 gallons. But at least one small U.S. company is working on plans for small bulk tanks. There are some small, very nice tanks available from Europe, but depending upon the strength (or weakness) of the U.S. dollar, these imported tanks can fluctuate greatly in cost. Small used tanks can still be found through used dairy equipment suppliers. These older tanks are often in varying states of disrepair and will probably require new refrigeration equipment and valves. (See appendix A for a list of suppliers of both new and used bulk tanks.) If you have a very small operation, there are some creative milk storage methods that many small creameries around the country are using successfully—with their inspector's approval. More on that coming up.

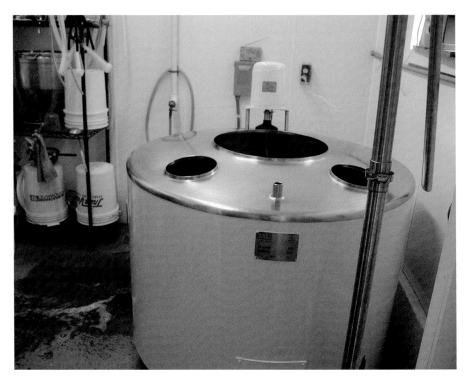
Small, rebuilt bulk tank, Twig Farm, Vermont.

A bulk tank consists of a stainless steel reservoir (sized to hold several days worth of milk) surrounded by an insulated double wall with compressor coils (to chill) built into the wall. Some older tanks utilize an ice bank (coils immersed in a water bath that create a layer of ice around the lines, providing chilled water that can circulate around the tank). In large dairies, milk is often pre-chilled in-line before entering the tank. This is more energy efficient and better for the milk, but it's usually not cost effective, equipment-wise, for the small dairy. Agitator paddles operated by a motor gently stir the milk to ensure even cooling as well as to prevent cream separation (important for cow's milk).

Important Bulk Tank Considerations
- **Volume of first milking.** When choosing a tank, it is important to determine the minimum volume required to attain a level in the tank that will ensure that the milk makes contact with the agitators (usually about 10%). You will need to know what your average production for one milking will be in order to determine if the tank size will work for you.
- **Storage volume.** Determine your milk production volume at peak month and maximum days stored. Try not to choose a tank that you will outgrow quickly.
- **Power supply.** Even most small bulk tanks will require a 220-volt outlet (think electric dryer or range outlet). Older tanks, and some that are specially made overseas, are often rated for 110 volts.

Creative Alternatives to a Traditional Bulk Tank
- **Combination chiller/vat.** Possibly the most practical solution for the small creamery is a vat and/or pasteurizer that is also supplied by a chilled water source for cooling the milk. Several companies are now outfitting units for this multipurpose use. A regular cheesemaking vat is outfitted to pipe in cold water from a remote chilling unit. This water is drained prior to cheesemaking, and hot water is then circulated in the vat. Using one piece of equipment instead of two or three can be the most economical and practical choice; however, be sure you deal with someone who knows the special needs of each function and isn't just putting together something that may compromise each process. Some considerations when purchasing a combination vat/chiller:
    - *Gentle agitation during chilling and cheesemaking*—variable-speed motor.
    - *Energy efficiency*—fully insulated exterior to maximize both the chilling and the heating efficiency of the unit.
- **Immersion cooling.** Pails or other acceptable containers are filled with filtered milk and then immersed in a cold-water bath. Milk is hand-agitated and temperature drop is documented either through a digital temperature logger or through manual logging. Some examples of ice-bath chillers include using a traditional bulk tank filled

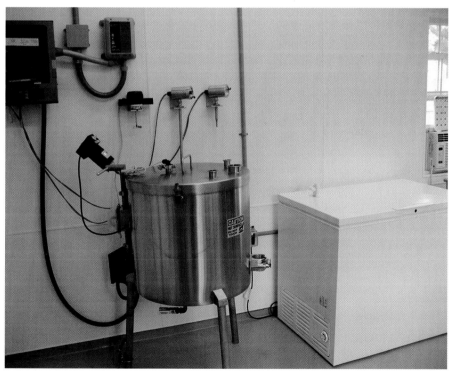

Pasteurizer/bulk tank combination chilled by homemade remote chilling system, Mama Terra Micro Creamery, Oregon.

with cold water into which food-grade bags of milk or milk cans are immersed and chest freezers filled with water, salt water, or a food-grade propylene glycol bath (both salt water and propylene glycol will attain very low temperatures without freezing) to cool cans of milk. Depending upon the temperature of the water bath, cans are either moved to a storage refrigerator or left in the cooling bath. This goes back to the old days when all dairies used milk cans and can coolers. Keep in mind that a 7-gallon milk can is very heavy when full, making lifting them in and out of a chiller physically difficult!

Bucket washer setup, Pholia Farm, Oregon.

Milk Line Washing Equipment

Whether you are installing a pipeline milking system or using a bucket milker, you may want to add a system that is specifically designed to aid in the proper and practical cleaning of your equipment. Lines can be effectively cleaned manually after each milking by using brushes, but this will erode the interior surfaces much more quickly than a vacuum washing system, as well as being very time consuming for you. The bucket washer (so called not because it washes your buckets—it doesn't—but because it is meant for use with bucket milker systems) consists of a suction unit that mounts to the wall onto which you attach your milking lines. A small sink or pail is filled with the washing solutions and the inflations are immersed in this pail. A line runs from the wall unit to your vacuum source. When the vacuum is turned on, the suction unit pulls and pushes the washing fluids through the lines. Between each cycle (rinse, wash, acid, or sanitizer) you will turn off the suction and change the solution in the small sink or pail. You hand-scrub the actual buckets and lids. A CIP system does

> **TIP**
>
> ### A Bucket Washer System with Faster Draining
>
> If purchasing a bucket washer system (look for one that has stainless steel components) that comes with a small side sink that hangs off one end of the wash sink, consider having a larger drain hole cut. Side sinks typically come with a very small hole in the center that drains rather slowly. Have your new hole positioned in the corner that is the lowest and closest to you (so the plug is easy to reach). Have it cut to the same size as the drain hole in your wash sink so you can use the same size plugs. Have a short section of stainless tubing welded to the bottom to direct the drainage downward (otherwise it will fan out and soak your legs!). This new, larger hole will greatly speed your cleanup time.

Pipeline milk hoses and clusters being washed, Sweet Home Farm, Alabama.

basically the same thing but forces the solutions through the built-in-place milk lines. As you might expect, a CIP system is more complicated and requires monitoring to ensure proper function. Consult with a qualified sales representative to make sure that your system is performing adequately and that you are using the correct detergents and sanitizers, water temperature, "slug" (air bubble) flow rate, volume, and time.

Milk Strainer and Receiving Pails
Whether milking by hand or by machine, the milk must be filtered. If you're using a pipeline system, an in-line filter will likely be used. If you're using a bucket milker, then you will pour the milk through a strainer/filter and into a receiving pail, or directly to a pail or tank if an in-line filter designed for a portable bucket system is used. Alternatively, a large filter can be placed in an opening on top of the bulk tank and the milk pipeline can drain into the filter reservoir. (*Note:* This works only if the bulk tank has an opening designed to hold a filter.)

Keeping Stainless Spotless
Be sure to inspect stainless steel equipment for cleanliness when dry-wetness will disguise residues and bio-films making dirty equipment look clean.

102 DESIGNING THE FARMSTEAD CREAMERY

Milk-can storage rack, Lazy Lady Farm, Vermont.

TABLE 8-3: Usage Guide for Common Dairy Chemicals					
Name	Common Chemical	Purpose	Water Temp.	Approx. pH or Concentration Goal	Cycle Run Time
Detergent	Potassium or sodium hydroxide (lye)	Dissolving of fat	120–160°F (dependent upon detergent and water)	Alkaline—pH12.0	Wash cycle 8–10 minutes
Chlorine Added to Wash Cycle	Chlorine	Dissolving of protein	Same as detergent	Alkaline; 50–100 ppm	Part of above cycle
Acid Rinse/ Sanitizer	Phosphoric acid, phosphoric-sulfuric acid blend, phosphoric-citric acid blend, hydrogen peroxide blend	Prevention of calcium deposits (milkstone); neutralizing of alkaline cleaners, acidic residue, for between milkings	70–110°F	pH2.0	2–5 minutes
Acid Wash	Same as for rinse, but with the addition of surfactants for additional mineral removal power	Used periodically after alkaline cleaner cycle	110–120°F	pH2.0	5 minutes
Sanitizing Pre-Rinse	Chlorine, iodine, quaternary ammonium, hydrogen peroxide	Pre-sanitizing of equipment and milk hoses	100°F	50–100 ppm (chlorine); 12.5–25 ppm (iodine); 200 ppm (quaternary ammonium)	30 seconds (no potable rinse needed)

When filtering into milk cans or directly into a bulk tank, choose a large, stainless steel strainer with disposable filters. These large strainers often come with a stainless steel screen filter—this filter is *not* PMO approved. You can order a second "punched" stainless filter that you can use along with a disposable filter. You will need to store disposable filters in a container that can be kept closed between uses. Special dispensers for filters can be used, or you can use any type of plastic or stainless container. We hung a small plastic trash can with a self-closing lid just above the sink.

Milk cans and receiving pails must meet sanitary standards. Rolled edges, rivets, rough welds, and glass lids are not acceptable. There are several inexpensive milk cans on the market for home dairy use—avoid these! Purchase cans that are listed as meeting "Grade A standards." When they arrive, inspect them for defects. Sometimes a rough weld can be fixed by fine sanding and polishing. Lids should be "umbrella" style, meaning they come up and over the outside edge of the pail (as opposed to nesting down into the neck of the can). Avoid rolled edges, as these will not be acceptable to most inspectors.

Other Stuff for Your Milkhouse

Milk testing equipment, such as scales, dippers, and milk measuring meters, can be stored in the milkhouse when not in use. This equipment is subject to inspection, so don't neglect to periodically take apart any milk meters for cleaning. Other items, such as kid/calf/lamb feeding equipment, might be allowed in the milkhouse, but check with your inspector. The philosophy should be: If it isn't used regularly, it shouldn't be in the room.

With thought and planning you can create a milking parlor and milkhouse that are efficient to work in and easy to maintain. Proper procedures in the parlor and milkhouse will help ensure a high-quality, safe milk supply—without which you negate any efforts to make good cheese! There is an old saying, *"Milk was never meant to see the light of day."* When you remove the milk from its source and send it through multiple processes, you put it at risk for becoming a dangerous or simply unappetizing food. Plan for and build spaces that help, not hinder, your goal of creating great farmstead cheese.

· 9 ·

Cheese Central: The Make Room

Now for the room that makes a dairy into a creamery, where science and art come together to transform highly perishable milk into a long-lasting, nourishing, and downright tasty product—cheese! The cheese make room is a place where you will be spending a great deal of time. It is important to make it not only highly functional, but also a pleasant place in which to work.

In this chapter we will discuss floor plan considerations, construction and maintenance standards, and equipment options. Appendix B contains sample floor plans from working creameries around the country. Feel free to refer to them during your study of this chapter to view placement options for the equipment mentioned. Appendix A provides a list of resources, including companies that supply much of the equipment described here.

Floor Plan Considerations

When designing the layout of your make room, you will need to address the following considerations:

- **Access** from other rooms, such as the milkhouse, office, and/or farm store.
- **Location** of nearby animal pens, dust, and other contaminants.
- **Size,** to provide adequate square footage for growth and additional equipment.
- **Windows,** to provide light and views.
- **Heating and cooling (conditioning)** that takes into consideration optimal cheese working temperatures.

Access
Convenient access from other workspaces and optimization of cleanliness should be your first priorities. Think about the proximity to your milkhouse for easy milk transport, either manually or by mechanical systems. Some of the best floor plans locate the make room centrally; in other words, you must first pass through other

Laini Fondiller hard at work in her small creamery, Lazy Lady Farm, Vermont.

rooms before entering the make room, progressing in levels of sanitation as you go. Remember, the farmstead creamery is located on a farm! An awareness of the constant challenge to keep dirt, flies, and pathogens out of critical areas will help ensure a safe, high-quality product. By locating the make room so that entry is through other areas, you will help minimize the introduction of unwanted pests and dirt.

I have never met a cheesemaker who complained of his or her make room being too spacious!

Size

Many people design their make room on the small side, picturing only one or two people working there and only minimal equipment in use. Don't forget about draining tables, racks, additional refrigeration units, etc. It is a good idea to try to plan a make room that is large enough to accommodate future, unforeseen needs.

Windows

A viewing window into the make room, either on the outside of the building or from your office or farm store, will allow guests and visitors to see into the room without actually entering. You may not be planning on having visitors, but there is a good chance that there will be authors, reporters, family members, retailers, and others who will be asking for tours. Another advantage to a window from the make room to the office or farm store is increased communication opportunities. If your operation is more than a one-person show, then it can be advantageous to be able to discuss such things as cheese orders with another member of the team while you are busy making cheese.

Position windows and exhaust vents away from animal pens and paddocks to help limit contamination of your workspace by dust, pests, etc. Consider installing filters in openings if such proximity is unavoidable. Also, at certain times of the year there may be more native molds and yeasts in the environment—this will affect your cheese! In larger creameries, "positive pressure" is often used to reduce their influence. (Positive pressure is when a room has more air pressure inside the room than outside. When you open a door to one of these rooms, you will feel a whoosh as the air flows out. This positive pressure keeps contaminants from entering the room.)

Heating and Cooling

When considering how to heat and cool your make room, first take solar (sun) exposure into account. Maintaining the correct temperature in your make room will be easier if you are able to anticipate the impact the heat from the sun will have on your space, see chapter 8. For example, our creamery is built of cement blocks (also called concrete masonry units, or CMUs). The make room gets a lot of late-afternoon sun exposure. In the winter this is fine, but in the summer we have to hang shade cloth to help keep the walls from accumulating too much heat (solar/thermal gain), and we keep room-darkening shades pulled down. By that time of the day, luckily, I am done making cheese, so I don't mind not having a view. We have deciduous shade trees and shrubs planted that will eventually help with this problem in the summer and still allow the walls to collect heat during the winter.

Optimal temperatures for draining most cheeses are right around 72ºF. If you are not able to maintain an optimal temperature through natural cooling and warming, you will need to provide mechanical methods of heating and cooling. Many small creameries use portable heaters and window air conditioners when needed. When we designed our creamery, we included radiant heat tubing in our slab (taking great care to keep it away from the aging room floor!).

Construction and Maintenance Standards

In the previous chapters of part 3, I referred frequently to the Pasteurized Milk Ordinance (PMO) when discussing construction and maintenance standards. It

> **DO-OVER STORY, PHOLIA FARM**
>
> There were very few creameries, ours included, that I visited where I did not see low spots in the floor from an improper sloping of the concrete. The best advice from everyone is to supervise the entire concrete pour and question any area that does not look to have adequate slope. We watched like hawks as the workers poured our floors—for the parlor, milkhouse, and aging room—and we thought they were doing well (they had "lots of restaurant floor experience"), so we left for errands. When we came back there was an obvious, very low spot against one wall. Too late to fix! Now we are stuck using a floor squeegee to get the water to the drain.

is important for you to remember that the PMO delineates standards for the production of "Grade A" products, such as milk and sour cream. As odd as it sounds, cheese is not considered a Grade A product; therefore, its manufacture is not discussed in the PMO. That being said, if your state follows the PMO guidelines for issuing your license, it will follow the same standards in the make room as it does in rooms like the milkhouse and milking parlor.

Floors

- Sloped to drain
- Concrete or impervious material
- In good repair

As with your other rooms, floors must be sloped to drain, made of concrete or other impervious material (as stated in the Pasteurized Milk Ordinance discussed earlier), and in good repair. "Good repair" is defined as free of excess cracks, chips, or pitting. Basically, ask yourself, "Can it be cleaned?" If a crack is likely to harbor residue and bits of curd, then it is not going to be acceptable. Epoxy-coated concrete (with texture), ceramic floor tile with well-sealed grout lines, or new concrete in good repair are all acceptable surfaces. If you go with epoxy, be sure to have a reputable installer do the work; many farmers have coated their own floors only to find it not adhering well after a short period of time. Creameries are very wet places that challenge the durability of most finishes. Even concrete will pit and erode over time from the acid of milk and sanitizers.

Walls

- Waterproof, cleanable surface
- Sealed junctions between floor and ceiling

Make sure your wall surface finish is durable and waterproof. This is one room where you cannot afford to shut down production in order to make improvements. (*Note:* If you are a seasonal producer, you can use your downtime for improvements and upgrades.) Surfaces such as epoxy-coated concrete block (CMUs), fiberglass-reinforced panels (FRP), dairy board, and painted plywood are all acceptable. When looking at the overall cost, consider the long-term maintenance. Junctions between the ceiling and wall, as well as around light fixtures, must be sealed in a manner so as not to allow dust or sediment to filter into the room.

Plumbing

- Adequately sized floor drains/floor sinks
- Hand-washing sink
- Double- or triple-compartment sink that is not trapped

You will need floor drains located and sized to accommodate floor wash water and water from untrapped sinks. Floor drains must also be accessible for cleaning and inspection. Remember to size the drains to handle a large volume of water without backing up. So-called floor sinks (they look like small, porcelain sinks with a large drain and are set down into the floor, with the concrete poured around them) are a better choice than standard shower-type floor drains. Some facilities have what is called a trench or gutter drain. These run the length of the floor and allow water to collect as it is shunted toward the main drain in the floor. Trench drains are acceptable but are not considered as sanitary as floor sinks: they allow for the possible accumulation of dirty water, increase the likelihood of splashback from sprayers, and provide a larger, hospitable surface for pathogen growth.

I recommend that any sink in which cheesemaking utensils and forms are washed not have a sanitary trap directly attached. (Sanitary traps are the curved piece of pipe located below household sinks. Traps keep a seal of water, situated in the curve, that blocks sewer gases from coming up the drain and out through the sink.) In food preparation areas, a sanitary trap can become a hazard, as it is another hospitable environment for pathogens. Any backup of water in the sink can bring that contamination into contact with your food contact surfaces. Instead of a sanitary trap at the sink, sinks can drain to a floor drain that is trapped. Hand-washing sinks can either be trapped or run to a floor drain.

In regard to floor drains and their traps, some jurisdictions will allow for the trap to be situated farther down the line (rather than directly below the drain). You will need to follow your local building code on this, but if traps are allowed farther from the drain itself, that is a more sanitary choice.

Plumbing in the make room, as in the other rooms, can run inside your walls (unseen) or on the surface. If it is on the surface, this is also an area that will need cleaning. Surface-mount plumbing is a good choice in such situations as our CMU construction or in retrofitting an existing building. Remember, if you surface-mount plumbing and live in a zone where freezing is likely, you will need to condition the building during times when below-freezing temperatures could be reached.

Electricity

You will most likely be working with a licensed electrician, unless you are competent to do your own electrical work and your jurisdiction allows for owner-builders to do their own wiring. While your electrician should be able to safely and

GREEN CONSIDERATION

We all know that compact fluorescent lights (CFLs) save a lot of electricity, right? Well, what most of us don't know is the potential contamination threat if one breaks. They contain small, but significant, amounts of mercury, which is a potent neurotoxin. If broken, this mercury is released into the environment. Go to www.energystar.gov for recommendations on cleaning up a broken CFL.

While it is still legal in most states to dispose of CFL bulbs in the trash, they should not end up in landfills. It is difficult to find local recycling options, but check these websites for help: www.epa.gov/bulbrecycling or www.earth911.org. Remember that because CFLs use so much less energy than incandescent bulbs (roughly 75 percent less), there is still a net environmental gain, as coal and other fossil fuel plants needed to make the power that typically supplies incandescent lighting are major emitters of mercury.

Extreme Green: If choosing CFLs, try to buy ones using technology which uses less mercury. Watch for cost decreases in LED (light-emitting diode) lighting, which has an even longer life and is more energy efficient than even CFLs—but LED lighting has a high up-front cost.

properly wire your building, there are some things that will be peculiar to the construction of a cheesemaking facility. Here are a few pointers you might want to share with your electrician:

- All switches and outlets will have to be "exterior" grade; that is, designed for wet locations (this applies to most of the rooms in your creamery). Outlets will all be ground fault circuit interrupter (GFCI) outlets, which are designed to protect you from electric shock in damp areas.
- Place outlets and switches above the 4-foot level on your walls (where it will be drier) and install them with wet location covers when situated near faucets, vats, steam, etc.
- Ample lighting over workspaces is required in most situations. Be sure to provide for 100-watt bulbs or the equivalent.
- Overhead lighting must be sealed and covered to prevent broken bulbs from contaminating product (see "Green Consideration" sidebar for more information on bulbs and safety).
- Install adequate mechanical ventilation. Commercial-grade exhaust fans are a good choice. Consider having them wired to a timer switch so they will automatically shut off after you leave the room.

Waterproof electrical cover, Mama Terra Micro Creamery, Oregon.

TABLE 9-1: Cheese Make Room Equipment Needs							
Cheese Type	Vat	Pasteurizer	Draining Table	Drying Area	Cheese Press	Aging Room	Brining Tank
Soft fresh, such as chèvre and fromage blanc	Optional***	Yes	Optional	No	No	No	No
Bloomy rind, such as Camembert types	Optional***	Probably*	Yes	Yes	No	Yes	No
Semi-hard, such as tomme	Yes	Optional**	Yes	No	No	Yes	Yes
Cheddar	Yes	Optional**	Yes	No	Yes	Yes	No
Parmesan types	Yes	Optional**	Yes	No	Yes	Yes	Yes

*It is possible to make bloomy rinds that are aged more than 60 days, therefore, from raw milk, but it is extremely difficult and their shelf life once sold is quite short.
**Can be made from raw milk only if aged 60 days or more.
***Soft fresh cheeses can be cultured in containers other than a vat if kept at the proper temperature.

Equipment and Accessories

The room in which you will make your cheese will have more equipment than any other room in your dairy. It is also the one in which you will probably be spending the majority of your time. As with the other room design choices, I highly suggest working in, or at the very least touring, several creameries before you settle upon your own floor plan and equipment choices. There is nothing like practical experience to help you define your own needs.

The following list contains common items found in most small, farmstead creameries. I have also included a chart to help you determine which main pieces of equipment you will need and under what circumstances.

- Cheese vat
- Pasteurizer
- Cheese press
- Draining table
- Drying racks/drying area or room
- Brining tank (can be located in aging room or make room)
- Antibiotic residue test kit
- Refrigerator/freezer
- Product refrigerator
- Rolling cart
- Double or triple sink
- Hand-washing sink
- Washdown hose

As we progress through this chapter, we will discuss each piece of equipment—choices, pros and cons, costs—and give detailed descriptions to help you sort it all out. Refer to appendix A for a list of resources for purchasing equipment.

Cheese Vat

A cheesemaking vat is a container in which you warm and culture the milk, coagulate it, and then "cook" the curd. A vat can be a very simple container, such as a pot, or an elaborate receptacle designed specifically for the purpose of warming, cutting, and stirring cheese curd. Cheese vats usually have a jacketed exterior where steam or hot water circulates to warm the milk. If you are making only soft, fresh pasteurized cheeses, then you may be able to use your pasteurizer instead, along with food-grade pails or tubs designed for the culturing of soft cheeses. Refer to table 9-2 for an overview of vat choices. Details on the different types follow.

Rectangular or Square Vats

Small rectangular or square vats are very uncommon in sizes under 100 gallons, although more companies are adding small processing equipment to their product lines. Rectangular vats have the biggest advantage when it comes to cutting curd, as regular cheese knives and harps fit nicely into the corners. When it comes to stirring the curd, they have a disadvantage—those same corners are not easily reached by automatic agitating paddles and will need manual stirring. Most rectangular vats are not covered, so during ripening and coagulation there is a disadvantage in keeping the top of the milk/curd at the proper temperature. Small rectangular or

Square cheese vat by Kleen-Flo, Twig Farm, Vermont.

TABLE 9-2: Cheese Vat Choices

Type	Price	Availability	Heat	Agitators	Curd Knives	Lid	Pros	Cons
Rectangular and Square	Low–moderate	Difficult to find in small sizes	Steam or hot water	Yes on larger units	Yes	No	Easy to cut curd evenly; handles larger volumes well	No lid means more difficult temperature control; curd collects in corners during stirring
Round	Moderate–high	Easiest to find new; rare used	Steam or hot water	Yes	Yes	Yes	Curd doesn't collect in corners during cooking; easier to clean; lid helps maintain warmth	More expensive than rectangular; cannot handle as large a volume
Combination Pasteurizer/Vat	Highest	Easiest to find new; rare used	Steam or hot water	Yes, but speed control issue	Possibly	Yes	Can pasteurize and serve as vat; same other pros as for round vat above	Most expensive; agitator speed sometimes too fast; often too deep for easy cleaning
Steam Kettle	Moderate–high new; low used	Easy to find new and used	Steam or hot water	No	No	Yes	Least expensive; more easily found used	Concave bottom makes regular curd knives ineffective; small sizes only (10–60 gallons)

square vats are difficult to find new and almost impossible to find used. New vats can cost from $6,000 up depending on size and extras.

Round Vats

Round vats are more readily available in a wide range of small sizes (25 to 200 gallons and up). Most include agitators and knives. Most also have lids (which help maintain temperature during ripening and coagulation). Most round vats have a tipping mechanism that slightly elevates one side of the vat so that whey will flow toward the drain more easily. Round vats are readily available from several manufacturers new; it is rare to find them used. If you are purchasing a new model, prices range from $12,000 to 20,000 depending on size and accessories, look for recommendations from other owners, as product and technical support are critical for initial setup.

TIP

Not All "Vats" Are Alike!

A "vat pasteurizer" is the same thing as a "batch pasteurizer." These units are not necessarily ready for serving as a cheesemaking vat! Be sure to clarify with the manufacturer that you need a vat in which you can make cheese.

Combination Vat and Pasteurizer

Combination vat and pasteurizers (sometimes called hybrids) are units that have been modified to perform both functions. The important thing to remember when considering a combo is that there is a difference between how milk is treated when it is being pasteurized as opposed to when it is being handled for cheese. For example, the speed at which a pasteurizer agitates the milk is too fast for stirring milk and curds when making cheese. So if you are buying a combination unit, be sure that modifications are included to properly perform both processes. Reputable manufacturers (refer to appendix A for equipment resources) will be able to assist you with these choices. Another thing to keep in mind is that even though any pasteurizer can be used to make cheese, some

> **UNIQUE AND RARE VATS**
>
> Believe it or not, there are a few copper cheese vats in use in the United States. They are fairly common in European cheesemaking situations, but inspectors in the U.S. have been quite resistant to their use here. Companies that have been able to use copper have had to convince the authorities that the type of cheese they wish to make cannot be duplicated without the influence of a copper vat.

200-gallon van 't Riet cheese vat (great reviews by cheesemakers on this Dutch company's service and products), with cheesemaker Alyce Birchenough at Sweet Home Farm, Alabama.

50-gallon round Qualtech (note mixed reviews by cheesemakers on this company's service) cheese vat and pasteurizer, Black Sheep Creamery, Washington.

are deeper than they are wide, which makes scooping curd, draining whey, and cleaning rather difficult. If you are planning on making only pasteurized-milk cheeses, then a combination unit is a wise investment. Prices range from $15,000 for a small unit up to $25,000 for a larger one. Even if you are making both pasteurized and raw-milk cheeses, though, it could still pay off. Compare prices—and remember, having only one piece of equipment means less space is wasted. Also compare energy usage. (See chapter 7 for more on calculating energy efficiency.)

Steam Kettles

Steam kettles, also known as soup kettles, are used primarily in restaurants for cooking and heating soups and stews. They are fairly expensive when purchased new, about $5,000–10,000, and, in my opinion, are not a good choice when compared to a new round vat, but they are more readily found used and often at a much cheaper price than a small used cheese vat. When shopping for a steam kettle, consider the following options:

- Tilting or stationary
    - Tilting will allow you to pour whey off the top
    - Stationary is less expensive
- Direct-plumbed or self-contained electric or gas
    - Direct-plumbed to a hot water or steam source in the room; cheaper than self-contained units
    - Self-contained electric creates its own steam; higher power bill
    - Self-contained gas creates its own steam; can be propane or natural gas

Watch your newspaper for restaurant closing sales, as well as classified ads in agricultural newspapers, and call restaurant salvage companies and supply compa-

TIP

Finding Used Equipment
Subscribe to dairy-focused "list serves" (email groups) on the Internet. There are many that are devoted to commercial dairy sheep, goats, and cows. Creameries that are going out of business frequently post their closures on these lists.

30-gallon steam kettle used as a cheese vat, Pholia Farm, Oregon.

nies that sell used equipment. If you are heating these appliances with hot water rather than steam, then safety concerns over buying a used unit are virtually nonexistent. There is little that can break on a direct steam unit; check that the drain valve has all of its parts and that any gaskets or O-rings are in good repair.

Pasteurizer

If you are going to make cheeses that require the milk to be pasteurized, then you will need either a vat/pasteurizer combo or a vat and a separate pasteurizer. Either way, this is an expensive purchase. Some people start out with very small pasteurizers only to find themselves processing several batches per day just to keep up with production (see Jan's story in the "Fraga Farm" sidebar). While this might save you money, your time is valuable too—so try to anticipate your production maximum for the first few years and install a unit that can handle that volume.

The PMO dedicates at least eighty pages of discussion to the proper use and maintenance of pasteurizers—including all the subjects you will be tested on. (You, or whoever in your creamery will be operating the pasteurizer, must pass a written and observed performance test before receiving a pasteurizer operating license.) I highly recommend referring directly to the PMO for study on this topic. Access this document online at the FDA's Center for Food Safety website (www.cfsan.fda.gov), or obtain a copy from your state's department of agriculture or other appropriate regulatory agency.

> ## FRAGA FARM
>
> Jan and Larry Neilson own Fraga Farm in Sweet Home, Oregon. Jan relates her first equipment experiences:
>
> "I bought the 15-gallon pasteurizer because I had no other choice; there just wasn't any other small equipment available and I wanted to get started. It was too small from the beginning, but I made two batches a day until I got the 30-gallon soup kettle, then I made up to three batches a day almost seven days a week. I got the soup kettle a year after licensing. After two months I needed bigger equipment. It was crazy since there were only two other producers in Oregon doing goat cheese. I started with 20 goats so I was pasteurizing like crazy. I had to increase the size of my herd within six months. There were a couple of times when I remember making six batches in one day!"

Prices for new pasteurizers range from $12,000 to $19,000 for very small units. The purchase of a pasteurizer is likely to be one of your most expensive budget items. This cost alone leads some creameries to choose to produce only aged, raw-milk cheeses. If you do decide to purchase a pasteurizer, choose the right size so you don't have to process more than one batch of milk per day!

HTST (high temperature short time) pasteurizers heat the milk to 161°F and hold it for only 15 seconds. These units are also known as plate pasteurizers or continuous plate pasteurizers (the milk flows through a series of heated and cooled plates for the processing). HTSTs are generally used only for high-volume needs. I mention them here only because you will see the name frequently in the PMO; when I use the term "pasteurizer" throughout this book, however, I will be referring to batch (or vat) pasteurizers.

TIP

PMO Standards for Pasteurizers

Remember: Any pasteurizer you purchase for cheesemaking must meet the PMO standards. "Calf milk," colostrum, and home pasteurizers do not meet PMO guidelines.

Batch/vat pasteurizers are used in small-volume settings, such as the farmstead creamery. They process milk by heating it to lower temperatures than the HTST units and then holding that temperature for longer. In addition to being more practical for low volumes, they are the most gentle on the chemical structure of milk. As I mentioned in the section on combo vat/pasteurizers, most pasteurizers can be used as cheesemaking vats, but they may require a few modifications. Also, some small pasteurizers are not shaped in a manner that would make using them as cheese vats very easy; for example, they may be too tall and deep, or too high on the legs, etc.

Pasteurizers are also very energy-demanding pieces of equipment. They require a 220-volt outlet, and even 15-gallon units will draw up to 40 amps. That means about 9 KwH per hour of use. A 30-gallon unit draws up to 60 amps, which means a usage of up to 13 KwH per hour. (See chapter 7 for more on calculating power usage and costs.)

Mechanical cheese press, Sweet Home Farm, Alabama.

Cheese Press

If you are making cheeses that require high-pressure pressing, such as Cheddar and/or Parmesan types (as well as many others), then you will need a means to apply up to 50 pounds per square inch (psi) to the cheese. There are vertical and horizontal pneumatic presses as well as mechanical Dutch-style presses; there are even presses that utilize water pressure. Again, searching for used equipment can be rewarding, but the choices are usually very limited. New presses for the small cheesemaker can range from around $1,500 to $4,000. You will also need forms (also known as hoops) that can be used in the press. Be sure the press and forms you buy will work for your volume and production goals. As mentioned before, you don't want to have to make more than one batch per day in order to use up your milk!

Draining Table

Many cheeses do not require the high pressure provided by a mechanical press. If you are making bloomy rinds, tomme styles, or blues, you will not need a mechanical press. These types can be hooped (put into the forms and molds) and then drained and/or pressed, if required (by weights, such as water jugs, barbells, etc.), on a draining table. A draining table has a stainless steel or food-grade plastic surface on which cheeses drain in the forms. It can be anything from a sloped stainless prep table to a custom-made draining table. It must be designed to allow

Pholia Farm's not-so-high-tech cheese pressing system.

the draining whey to flow away from your pressing and draining cheeses and to be collected in pails. If you are making only soft fresh cheeses, then you will not need a draining table, but you will need a draining rack—something to hang your draining bags from and a way to collect the whey below them.

We happened to find a stainless steel dish-draining unit at a used restaurant supply sale. We had stainless legs welded to it at a height that was comfortable for me to work (which is a bit higher than most would like!). It is working well for us and was very reasonable in cost.

Draining Racks

If you are making soft, bag-drained cheeses, such as chèvre or feta, you will need a sturdy draining rack or rod from which to suspend your cheesecloth bags over tubs to collect the whey. This is another area where you can improvise. Keep in mind the usual criteria of cleanliness and optimal temperature. Try to keep the area situated away from drafts, heaters, etc.

Drying Racks/Drying Room

Drying racks are needed when you produce bloomy-rind (also known as surface-ripened) cheeses, such as Camembert or Crottin. These can be purchased through cheese-supply companies. Choose the type of food-grade matting that will best suit your finished product goals. The more difficult issue is where to place the drying racks, as the bloomy-rind cheeses need a different environment than is usually present in most make rooms, aging rooms, or coolers. Be sure to take this need into consideration when designing your space. If you are not confident of the temperature and humidity needs of your particular cheeses, consider hiring a consultant, taking courses pertinent to your cheese types, and studying cheesemaking guides (see appendix A for resource lists).

Brine Tank

Brine, or brining, tanks are used in the production of many aged cheeses. Some cheeses, however, are dry-salted (as curd or wheels). If you are making a cheese that requires brining, you will keep a tank or reservoir of brine in your aging or drying room, or you will have a freestanding tank that is cooled (brine should be 50–55ºF) by an external source of cold water (like what is used to cool a pasteurizer). Depending upon your production volume, something as small as food-grade covered tubs available from restaurant supply houses, meant for storing foods like salads, can be used. You can also purchase brine tanks specifically designed for cheese production at some of the suppliers listed in appendix A. Prices will vary according to size and materials.

> ### HOW CHEDDAR WON THE WEST
>
> Historically, Cheddar has been the cheese of choice for most Americans, closely followed by mozzarella and Jack cheese. None of these cheeses are brined—ever wonder why? Respected technical cheese expert Neville McNaughton observed the following in a *Cheese Reporter* article:
>
> "Brining has traditionally required a great deal of space, which adds to the capital cost of a cheese production facility. It also involves additional time and labor; brining a 20-pound wheel may take three to four days, a 3-pound ball-shaped cheese about 30 hours. Cheddar is in the block in about five hours or less."
>
> Count on Americans for choosing expediency!

Antibiotic Residue Test Kit

Many states are now requiring that even farmstead producers who milk only their own animals (and yes, even certified organic producers who cannot regularly use antibiotics and remain certified) to perform an approved drug residue test on every batch of milk they make into cheese (this is part of appendix N of the PMO). For many small cheesemakers, this seems both an unnecessary cost as well as a poor use of their time, but it is becoming a widely enforced law. Before being allowed to begin production, you will receive training and a test from the governing agency and its representative in the appropriate department. In addition, your regular inspector will be checking your records during his or her unannounced visits. For most small producers, the procedure sounds intimidating, but it is relatively simple.

The most common test kit for the small

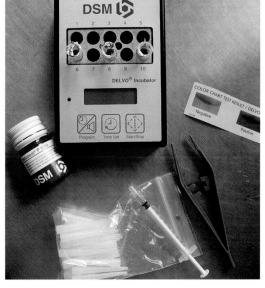

Delvotest P Mini antibiotic residue test setup with incubator.

producer is the Delvotest P mini. In addition to the incubator and test kits, you will also have to purchase "positive controls." This is penicillin in either tablet or vial form that is at a specific concentration meant to be the lowest amount detectable by the test.

Contact your governing agency for information on learning the specifics of your test's requirements. Tests other than the Delvotest P may be required or suggested. Calibration requirements, thermometer requirements for tracking temperature during storage of the test vials, as well as the area in which you will be allowed to set up your "lab" may be dependent upon rules particular to your state. You would be wise to research this early in your planning to avoid any difficulty later on.

Refrigerator/Freezer

You will need a dedicated refrigerator for storage of rennet/coagulants, cheese that is ready for sale, ingredients used in cheese, and antibiotic residue testing kits and positive controls. A dedicated freezer will be needed for cheese cultures and ice packs if these are being used for shipping cheese. ("Dedicated" means that you cannot store your leftover dinner or any other personal snacks in your cheese-supply refrigerator or freezer!) You can use any refrigerator—household or commercial—as long as it is in good condition and is cleanable. If the freezer has an automatic defrost, it may not be appropriate for storing starter cultures; your source for purchasing culture will be able to advise you as to its storage parameters. You will need at least one refrigerator to store product that is ready for sale and wheels that have been taken out of aging and cut. It is also useful to have enough refrigerator space to pre-cool wheels for shipping. If you are storing your antibiotic test kits and positive controls in this refrigerator/freezer, then you will need to have approved thermometers (usually wet-bulb, calibrated in Celsius)—one on the top shelf, one on the lower shelf, and one in the freezer compartment. You will probably want a chart on the outside of the unit on which you will record the daily temperatures (part of the antibiotic residue test requirements).

Rolling Cart

While a cart may not seem like a piece of equipment that should be at the top of your priority list, I have found ours to be one of the single most versatile and often-used items in our creamery. From moving full cans of milk, wheeling cheeses out of the aging room for affinage (a French term for the craft of maturing and aging cheeses), to simply serving as an extra surface when needed, our rolling cart is invaluable. We chose a composite plastic cart, such as those made by the company Rubbermaid, instead of a more expensive stainless steel version. Whichever type you choose, be sure that it will fit between shelving in the aging room and that it is easy to maneuver.

Triple-Compartment Sink

A triple-compartment sink has dedicated wash, rinse, and sanitize compartments. You may be allowed to use a double sink if you are going to pre-sanitize your

Sump pump used to remove whey, Sweet Home Farm, Alabama.

equipment just before use, but I recommend installing a triple sink even if your inspector approves a double—the convenience will pay off over time. Consider not trapping the sink drain, instead allow it to drain into the floor drain (which is trapped). This is a safety feature to prevent contaminants from migrating from your trap into your sink. Triple sinks can be square or curved, and with or without drainboards. For ease of cleaning include both a pre-rinse sprayer and a fill faucet.

Shopping around for used triple sinks can save quite a bit of money. New, they can cost up to and even more than $1,000; used, they can be found for only a few hundred dollars. Another factor to consider is the gauge (thickness) of steel used to manufacture the sink. Many of the newer sinks are only 22-gauge steel (the lower the gauge, the thicker the metal), so they are rather flimsy and easily dented, and they are so thin it can even be difficult to mount them straight. Look for 16- and 18-gauge steel. Often the older used sinks are of heavier gauge. You may have to clean off a lot of milkstone and other residues, but it will probably be worth it. As with the other sinks for your creamery, watch trade papers, newspaper ads, and the like for restaurants going out of business; also check salvage shops.

Hand-Washing Sink

For the make room, if you can find a used foot-pedal-operated sink, or spring for a new one, you will be glad you did. This is the room in which you will want to

> ### DISHWASHERS AND CLEAN-OUT-OF-PLACE (COP) WASH SINKS
>
> Many cheesemakers utilize commercial dishwashers or a COP sink in their make room for washing utensils, forms, and more. Commercial dishwashers use high heat as well as chemicals to very rapidly clean and sanitize (about 90 seconds per load). They are expensive, but can often be found used at larger restaurant supply stores or in industrial classified ads. Walter Nicolau, of Nicolau Farms in Modesto, California, says he could not live without his commercial dishwasher.
>
> COP sinks are basically large tanks in which dirty utensils and equipment are placed, then water (with cleaning chemicals) is pumped into and through the tank. Alyce Birchenough, of Sweet Home Farm in Alabama, tried out a COP sink at their 13-cow dairy and found that although it worked very well, it was oversized for their operation, which produces about 13,000 pounds of pressed cheese per year. Alyce prefers their under-counter commercial dishwasher; however, she reports that cheesemakers who produce mostly soft-ripened cheeses, necessitating the use of many small plastic forms, find the COP tank very worthwhile.

be the most sanitary. A foot- or knee-pedal sink will take one more critical step out of the process. Remember, your hand-washing sinks can either be trapped or drain to the floor drain.

> **TIP**
>
> ### Mounting a Curved-Bottom Sink
> If you install a curved-bottom sink, I recommend buying a set of wall-mount brackets instead of legs. This makes it much easier to clean under the sink and leaves fewer things to stub your toes on!

Washdown Hose

Have the washdown hose faucet in your make room installed in a central location so you can easily clean the entire floor of the room. Make sure to provide a hose that is rated for hot water; these are usually a dull red color. Sprayer heads should be able to withstand hot water usage, as well.

Other Stuff for Your Make Room

There are a few additional items you might want to consider to help keep your make room more organized, efficient, and comfortable.

Tables

You will need at least one stainless steel or other food-grade surfaced prep table for packaging product. Regulations in your area may also allow you to do your

> **TIP**
>
> ### PVC—The Cheesemaker's Friend!
> I have seen many useful things in make rooms that have been cleverly constructed of PVC pipe—shelves, rolling sink carts, cheese hoop holders, platforms to elevate refrigerators off the floor, and more. PVC is easy to work with, inexpensive, strong, and, best of all, water- and rust-proof. Keep it in mind for applications where stainless steel is not necessary!

Cleverly constructed cheese form drying racks made of PVC, Sweet Home Farm, Alabama.

antibiotic testing on the same surface; if not, you will want a separate table or area for lab and testing equipment. Another small table for taking notes and maintaining your product make sheets and inventory logs will also be handy. You can often find used stainless steel prep tables at restaurant supply stores or through industry-related classified ads.

Shelving
Above your packaging table, you may want some shelving for frequently used packing materials such as plastic wrap, labels, and freezer paper. You will not be able to store packing materials, such as boxes, in the creamery because they will absorb humidity and mildew, becoming a potential source of contaminants. Shelving above your sink can store your clean cheese molds (hoops/forms) and utensils. Shelving for detergent and sanitizers should be located away from and/ or below anything having contact with product. Remember, chlorine will erode

TIP

Check Your Shelving Hardware!
The hardware used to anchor wall shelves will eventually rust and corrode. Checking it once a year or so might save you from a falling shelf and damaged equipment!

stainless steel, so choose shelving of other material that is easily cleaned and noncorrosive (meaning it won't corrode in the presence of most chemicals). Plastic-coated wire is a good choice, but it will also eventually show signs of rust where the plastic coating erodes or the metal is exposed.

The cheese make room is the most complicated space in your dairy, and the choices you will need to make in equipping it properly can be daunting. Because each creamery has its own unique production needs, there is no single blueprint for setting it up perfectly—often, no amount of proper planning will likely fully anticipate the direction your company will take in regard to product lines and volume. Don't worry! If you have created a sound business plan, you will have a good idea of the starting needs of your business and you will be able to outfit a workable make room that will serve you well.

· 10 ·

Aging Rooms, Cellars, and Caves

Nothing quite conjures up a romantic, gastronomic response when talking about fine cheese like the term "cave-aged." The consumer's mind is immediately transported to a cool, limestone vault where wheels of cheese sit aging on wooden planks—even if the reality is an aboveground walk-in cooler with wire racks and a complicated refrigeration system. While I personally have a bit of an issue in referring to such a system as a "cave," the term is becoming accepted usage in describing any space dedicated to aging cheese. (It's similar to the practice of calling a wine storage room a "cellar" even if it is not underground.) For this book, however, I will refer to any aboveground aging space as an aging room and any underground space (created by humans) as a cellar or cave. Call it what you will, the proper building of a functional aging space can add both tangible and intangible value to your cheese.

The type of cheese you choose to make will dictate the design of the aging room. Some cheeses will also require a drying room for the first phase of aging. (Drying rooms are similar to aging rooms but usually have a lower humidity and greater air flow.) The drying room can also contain brine tanks (instead of the aging room). I will not talk about drying rooms as a separate topic in this chapter, as the design issues and equipment will be the same; only the application will differ.

I will approach the topic first by discussing the "big three" issues facing the

AGING ROOM RESOURCES

A very fine, in-depth study of aging room options was done by Jennifer Betancourt with help from Amanda DesRoberts in their SARE Farmer Grower Grant publication *Current Options in Cheese Aging Caves: An Evaluation, Comparison, and Feasibility Study*. You can access it on Jennifer's farm website at www.silverymooncheese.com or via a web search. A lot of thought and work was put into this publication, and it is well worth studying. Also, Jim Wallace, with New England Cheesemaking Supply, and private consultant Peter Dixon have both documented many of the steps needed for cave and aging room construction—Jim on the company's website (www.cheesemaking.com; click on Aging Help in the Help section) and Peter through his own publication *Farmstead Cheesemaking* (available in back issues) and his writing in books and for the periodical *CreamLine*.

affineur (cheese-ager): temperature, humidity, and air flow/movement. Then we will go over design, equipment, and construction options. Finally, we will discuss some unmentionables, such as cheese mites, that might be a future obstacle for you and your cheese.

The Aging Room: The "Big Three" Design Issues

When designing an aging space, there are three main factors to consider: temperature, humidity, and air (both air exchange and movement). Each of these factors will be influenced by the volume and type of cheese being aged; your area's native influences (weather and soil/earth stability and properties; also, if your building is underground); and the building structure and size. Because of the uniqueness of each building situation, there is no single book or guide that will be able to accurately tell you the perfect design for your aging space. By understanding the requirements for a good aging room, however, you will be able to evaluate the options for your space and successfully design a satisfactory aging room.

> **By understanding the requirements for a good aging room, you will be able to evaluate the options for your space and successfully design a satisfactory aging environment.**

Temperature

Most cheeses age beautifully at between 50° and 55°F (10° and 12.7°C). Certain cheeses, however, such as bloomy-rind types (e.g., Camembert and Brie) and Emmentalers (eye-formation cheeses), will have unique needs at different stages in their maturation. It is obviously very important to know what type of cheese will be inhabiting your aging space. Quite often cheesemakers will have more than one aging room, even if one is a walk-in at 55°F (12.7°C) and one a commercial refrigerator at 38°F. While many people simply plan on using mechanical, power-thirsty compressors to cool their space, if you plan well you will be able to greatly reduce, and perhaps eliminate, the need for equipment that will affect your profit margin by raising energy usage costs. I have written each of these sections under the assumption (and hope) that cheesemakers will attempt to design a system that is as energy efficient and environmentally friendly as possible.

Often an aging space is constructed that can maintain a constant, year-round temperature in the acceptable range for aging cheese. But then, once that room is filled with cheeses, an additional cooling source must be added. The fact that aging cheese actually generates heat is often overlooked. The enzymatic processes occurring within aging cheese cause the release of energy, thus creating heat—which is all fascinating, but what does it mean for you as the affineur? According to author Jean-Claude Le Jaouen in his book *The Fabrication of Farmstead Goat*

Young cheeses aging at Consider Bardwell Farm, Vermont.

Cheese, one ton of ripening cheese will release between 1,000 and 3,000 calories of heat every 24 hours. Just what is a calorie? One calorie is what is required to raise one liter of water by one degree centigrade (1.8°F). So let's say you have aging space for two hundred 10-pound wheels (that is exactly one ton of cheese). Even though it won't all be going through the same stage of enzymatic activity at one time, let's assume that it is releasing 3,000 calories every day. You can see how, over time, you will have a temperature increase in your aging space that could be detrimental to your cheese. So even if your structure is able to maintain a constant 55°F temperature when it is empty, you will not be able to hold that temperature without some sort of cooling assistance once the room is filled with cheeses. Remember to do your calculations based on the maximum amount of cheese your room can hold; otherwise you will be in for an unpleasant surprise at the worst possible time—when you are at maximum capacity.

The fact that aging cheese actually generates heat is often overlooked.

Several other factors can influence your aging space's ability to maintain an optimal temperature without excessive need for mechanical cooling. These include temperature fluctuations for your area (especially night-to-daytime differentials); soil thermal potentials, in both R-value (insulation capability; see sidebar) and thermal mass (heat-storing capability); site location (north-facing being ideal); and even shading from trees and cooling masses of green surface plantings.

You can use the Internet to search for temperature records for your area. Finding out average highs and lows (both daytime and nighttime), as well as record highs and lows, will help you anticipate the possible temperature fluctuations and plan for them adequately. Sustained highs or lows without a nighttime variation will make it more difficult to recover an appropriate temperature for your cheese. For example, if your daytime summer high is 98°F (like it gets where we live) but your nights cool off to the mid-50s, you have a good chance of storing some of that coolness in thermal mass in your building. But if that same thermal mass is exposed to the hot sun all day (say your aging room faces south), then it will be more difficult to maintain an adequate aging temperature. (More about thermal mass later.)

If you are building an underground aging space, the material you cover it with will affect your temperature control. Soil, according to one source, has an average R-value of 0.25 per inch. (Solid granite has an R-value of 1.0 per inch.) So 5 feet of soil covering your aging cave will give you a value of about R-19. But soil's other property of thermal mass will allow you to hold temperatures for longer (traditional fluffy insulation, by contrast, has a very low thermal mass). This can work for you, or against you, depending on what is happening that might heat up the soil or rock. For example, I have seen some beautiful aging cellars that are faced with stone, but when exposed to the sun all day, the stone heats up and transfers much of that heat to the structure behind it. This is why it is ideal to face the aging space north (no direct sun) and/or plant shade trees or vines to help keep that thermal mass from building up too much heat during the hotter parts of the year. You can also use these properties to help warm a too-cold cellar during the winter, by making sure that those shade trees and vines are deciduous—when they lose their leaves in the winter, the rock and soil will be exposed to the warmth of the sun. This is why it is so important to know the temperature extremes for your area before you decide upon the material and orientation of your cheese cave.

> **R-VALUE EXPLAINED**
>
> R-value is the ability of a material to resist the transfer of heat (or cold, for that matter). A single layer of window glass has an R-value of 1. Most homes rely upon high-R-value insulation to prevent heat loss through walls, attic, and floor space. Wood-framed (stick) structures have insulation blankets (batts) placed between the wood framing members (or loose insulation fill pumped into the open bays between the wood). While this insulates the open space well, each wood framing member will have a much lower R-value, allowing for heat loss and gain in an otherwise well-insulated wall.

Nothing says your aging space has to be in a separate building. Many fine aging rooms are attached to the cheesemaking room or to a farm store, or located underneath the main creamery. Whatever your approach to the space, you will want to optimize the materials selected as well as create enough space to house the maximum amount of cheese you plan on aging at any one time.

Volume of product is the main concern when determining the size of the aging space. Not only will you need room for the cheeses, but you must leave enough space behind and between shelves as well as between cheeses to allow for air flow. Large wheels will optimize space better than small ones. So before you decide on the size of the room, sit down and figure out your approximate volume of cheese (in pounds or wheels) and how many shelves you will need. Then you should be able to determine what size room you will need to accommodate that volume. Be sure to allow for some extra shelf space for times when production is high or outflow of product is low. Also, don't forget to factor in room for brine tanks!

When constructing your aging room, it is almost impossible to build it with too much insulation (there's that "R-value" again). In residential "conditioned" spaces (meaning spaces that have heating and cooling controls), most heat loss is through poorly insulated ceilings, leaks around windows and doors, and thermal mass that works against the desired temperature goals (such as concrete slabs).

Here is a perfect example of thermal mass working against the cheesemaker's goals: We thought we had planned out our first aging room very well. We situated it totally interior (meaning none of its walls were exposed to the outside), did not run any of the hydronics (in-slab hot water heating system) close to the space, filled the concrete block walls with pelleted insulation, and provided a wine cellar cooling unit sized properly for the space. It worked great until the really hot part of summer hit—over 100°F for several days. Then we watched in horror as the cooling unit worked nonstop and the temperature continued to creep up. Vern took his handheld infrared temperature sensor and aimed it on the walls and floors. Even though the slab in the aging room was not directly a part of the slab outside the room, the sensor's probe clearly showed the cool leaking out of the room and the heat creeping into the room via the slab. So the lesson is, don't discount any building material that will act as a reservoir for heat or cold. Choose materials appropriately to work for you, not against you. (More on building material options later in the chapter.)

So what did we do to fix our little problem? We waited until the following winter, moved all of the cheese out of the room, and basically made the room smaller. We added a layer of rigid foam insulation (with an R-value of 14) to all of the walls and covered that with fiberglass-reinforced panels (FRPs). We didn't change the floor or ceiling, figuring we could do that later if this was inadequate. Now the aging room cooler runs only during the hottest of days. As the wine cellar cooler we use to chill our aging room vents directly into the cheese make room, we also have a window air conditioner that we run during those same time periods. If this room gets too hot, then the cooling unit cannot function at its optimum level.

Other factors, such as heat from lights, fans, and even your body, will add to the temperature control challenges of an aging room. Using cool bulbs, such as compact fluorescents, and limiting your time in the aging room will help. (For example, putting your cheeses on a cart and taking them to a separate room for rind care, instead of performing that work in the aging room, will help keep heat contributions from your body to a minimum.)

Humidity

Relative humidity (referred to as simply "humidity" throughout this book) is as important to cheese aging as temperature, but easier to control. As noted with temperature, humidity goals will vary depending upon cheese type, but most will require humidity levels between 85 and 95 percent.

Without the proper level of moisture (called available water), the bacterial and enzymatic processes that must occur during cheese aging will come to a screeching halt and your cheese will, to put it bluntly, die. So the cheesemaker who is also

RELATIVE HUMIDITY (RH) EXPLAINED

The amount of water vapor that air can hold at a specific temperature is called relative humidity. When air reaches 100 percent humidity, it is fully saturated and condensation will occur. Warm air can hold more moisture than cold air. For example: If your aging room is 55°F and 85 percent RH and you decrease the temperature, the RH will go *up* because the amount of moisture the cool air can hold is lower. (Here is an experiment I like that demonstrates relative humidity: Take an airtight container at room temperature and normal humidity, close it tightly, then place it in a freezer. After a few hours, open the container—you will find that ice crystals have frozen to the lid. The colder air could not hold the moisture and it condensed out as ice crystals.) This is why it is typically easier to maintain humidity (meaning RH), all other factors being equal, in cooler climates. (Not to mention that when refrigeration equipment runs it removes moisture from the air.)

the affineur must know the ideal humidity required for his or her cheese type. For some cheeses this will not be the same throughout the entire aging process, so either you will need a way to change the humidity, or you will need two separate aging spaces.

As with weather patterns, you should be able to look up humidity ranges for your geographical location. Some of us live in areas where winters are cold and dry, others where winters are cold and wet. Knowing the seasonal variations for your area will help you anticipate the needs for your space.

The same features of a building's structure that aid with temperature control will also assist with humidity control. A well-insulated building that does not have a temperature difference between the interior surfaces and the air in the space will not likely have condensation problems on the walls and ceilings. I will talk more about this in the equipment section, but just think of a glass of ice water in a warm room with droplets of condensation beading up on the glass—you don't want this to occur in your aging room! Water in the *air* is good—but not on the walls and ceiling where it could drip on your cheeses. Aging rooms built under living spaces (which are kept warmer than the aging space) and not adequately insulated will have this problem, as will cellars where ice water is pumped through radiant tubing in the walls. The material used to construct the walls and floor can have a great impact on relative humidity in the aging room. Brick walls are probably the ideal, as brick has the best potential for absorbing and releasing moisture without having condensation issues. Brick structures are much more readily accepted in Europe, but you might be able to at least make a case for a brick-paved floor in your aging space.

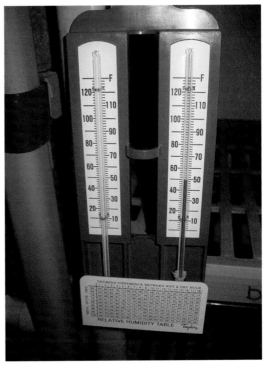

A psychrometer for measuring relative humidity by comparing the difference between readings from wet and dry bulb thermometers.

Water in the *air* is good—but not on the walls and ceiling where it could drip on your cheeses.

Air

Cheese needs both air movement and air exchange to age properly. I have tasted (and smelled) cheeses at competitions that were obviously not aged in adequately ventilated spaces. Musty cellar odors are not what your cheese should exude. Remember to think in terms of both the rate of flow (movement of air across the cheeses) and the volume of exchange (how often the entire air volume in the room is replaced with fresh air). Of course, the more air you exchange, the more challenged you will be to maintain steady temperature and humidity!

Some cheeses need more oxygen in the beginning stages of ripening; others produce ammonia gas and carbon dioxide during ripening. These cheese types, especially when you have a large volume in proportion to the space, will require some means to exchange the "used" air with fresh, clean air.

Air quality and movement are not the only factors you need to keep in mind for the aging room—you also need to think about what is *in* that air. There are certain molds and bacteria that cheesemakers want to encourage in the aging room space; however, there may be times of the year when you will have to deal with unwanted spores entering the fresh air supply. Very few of us will be able to build a totally controlled, airtight, positive-pressure aging room. But never fear, this does not mean the demise of the cheese, only the unlikelihood of producing a uniform, possibly uninspired product. Most artisan cheesemakers learn how to deal with unwanted surface molds through various affinage techniques.

An oversized room will have fewer air issues than a small, overcrowded aging room. There must be adequate airspace both behind shelving as well as between cheeses to allow for air circulation. For optimal, natural air movement in the aging room, vaulted ceilings are the best (although most costly) choice. Locate shelving several inches from the wall surface to facilitate airflow along the walls and also to keep cheeses away from any moisture that might accumulate on wall surfaces. If adequate air movement is not achieved (placing thermometers at different levels in the room will help you determine this), then a small fan (not blowing directly on the cheeses) can assist with air movement. Again, the more cheese you have in the room, the more impedance there will be to air movement—so be sure to monitor the stratification of air in the room as your aging room fills with cheese.

Design, Equipment, and Construction

Now that you understand the three most important issues that affect your aging space—and what you can do about them—you can more readily choose the other options and features of the aging space, such as floor plan, cooling (and possibly heating) options, humidifiers, shelving choices, and construction material options.

Floor Plan and Utilities

The single most overlooked floor plan feature in many cheese cellars and caves is an antechamber—a room you enter before you enter the aging space. The antechamber/entry room (we won't call it a mudroom!) will serve first and foremost as an airlock and thermal barrier that will help keep the aging room from experiencing fluctuations in temperature and humidity. The best antechambers have

> ### CEILINGS: TO VAULT OR NOT TO VAULT?
>
> Most cheesemakers and affineurs will agree that, all other things being equal, a vaulted ceiling is preferable to a flat ceiling. This type of configuration has two main advantages: Airflow is improved, by allowing for a natural, circular flow; and any condensation on the ceiling is less likely to drip onto cheeses but will instead run to the sides and down the walls. Then there is that romance factor.
>
> Along with wood shelves, a vaulted ceiling in your cave or cellar oozes old-world charm. On the downside, a vaulted ceiling will increase the cost of construction considerably. I would suggest investigating the options for vaulting, while keeping in mind that plenty of wonderful aging spaces have flat ceilings, not dripping with moisture, out of which have come some of the most acclaimed aged cheeses in the U.S.

at least one sink, a place to change shoes and hang lab coats, and temperature and humidity charts for documenting proper aging temperatures (part of a good quality assurance program; more on that in chapter 12); in addition, they offer access to any mechanical systems necessary for your aging room. Many people also install a window between the antechamber and the aging room, to allow visitors a peek at the cheese without having to enter what should be one of your most protected rooms. When I interviewed people who have built caves and cellars that are not attached to their cheesemaking facility, the number-one thing they would change about the design was to add an entry room and provide a sink for washing, shelving, etc.

When considering lighting for the aging room, choose vapor-resistant, shatterproof fixtures (just as you did for the milkhouse and make room). Also remember that lighting will create heat, so whenever possible, select compact fluorescent or other "cool" light bulbs.

In the main aging room, calculate space for shelving, including a gap along the wall, as well as comfortable working space between shelves. If you will be transporting your cheese to and from the room on a cart, make sure you have enough space to maneuver without running into shelves. And don't forget to provide space for brine tanks, unless you will be using a cooled, freestanding brine tank in your make room.

Equipment

Now that you have an idea of how you want your aging room space to look and feel, what equipment will you need to put in it? While we introduced you to the importance of the "big three" design issues earlier—temperature, humidity, and air—let's look at some items that will help you address these issues and prepare your aging space for ripening cheese!

Coolers

You will have several choices for mechanical cooling assistance, should you need it for your aging space. Typical choices include traditional refrigeration units (often referred to as condenser/compressor units), floral cooling units, Burch Industries

> ### "LOW-VELOCITY" COOLING
>
> The key phrase with any of these cooling units is "low velocity." In a nutshell, it is the velocity of the air moving over the cooling fins, or coils, combined with the compressor's capacity that dictates the amount of humidity removed from the air. Maybe a bit too technical for most of us (me included), but when your heating, ventilation, and air conditioning (HVAC) specialist starts spewing information, perhaps the glaze-over will be minimal! Suffice it to say that any system not specifically designed to cool and condition cheese—as opposed to something else—will have technical issues to overcome.

egg room coolers, wine cellar cooling units, chilled water lines, and window air conditioners with a CoolBot (more on those later). The important issue with any unit is customizing and optimizing it to meet your cheeses' needs without excess energy loss/waste and/or undue financial investment.

Refrigeration units are designed to—guess what?—refrigerate, so if you are aging at refrigeration temperatures as well as waxing, vacuum-sealing, or otherwise protecting your cheese from drying, this type of cooling will work well. But if you are aging at warmer temperatures, 55°F being the most common, then modifications will be required to maintain proper temperature and humidity. Refrigeration equipment should be selected with the assistance of a qualified heating, ventilation, and air conditioning (HVAC) specialist or manufacturer of appropriate units.

Egg coolers (the only U.S. manufacturer is Burch Industries; see appendix A for contact information) are similar in look to condenser/compressor refrigeration equipment except that they are designed to keep eggs at either 50°F and up or 45°F and up (depending on whether the eggs are intended for hatching or for table use). They can also be equipped with humidifiers and heaters. The smallest units will cool 924 cubic feet.

Wine cellar cooling units are another good option. They come in many different sizes, keep the cheese at the proper temperature, and remove minimal humidity from the air. The humidity goal for wine aging is 50 to 75 percent, but the moisture removed can usually be drained directly back into the room, depending upon the unit's format (self-contained or split). Wine cellar coolers are quiet, efficient, and easy to install. The biggest drawback, as far as I can determine, is that the self-contained units must vent the heat removed from the aging space into a room whose ambient temperature does not exceed 80°F—in other words, a conditioned room. Otherwise the unit will have to work too hard to cool air as it goes into the aging space.

Chilled water lines can be plumbed as a part of the walls of the aging room. I have seen this done by cutting channels into rigid insulation and embedding flexible water tubing (known as PEX), and even by chiseling channels into cement block and placing the water lines in the grooves; both were then covered with a waterproof surface. Chilled water (from an ice bank or remote chiller) is then pumped through the tubing, thus cooling the walls and radiating into the aging space. The wall behind the piping should be covered with reflective insulation so as to minimize both loss of cooling to the outside as well as heat gain from outside of the room. The biggest drawback I have seen for this type of system is the accumulation of condensation on the walls. If you avoid putting the cooling pipes in the ceiling (where the condensation could drip onto cheeses), this should not be too big of an issue.

The use of *window air conditioners and a CoolBot*, a relatively new development, is an interesting option for cooling your aging room. Residential air conditioners are not designed—and indeed have no controls—to cool air to below 65°F, but the CoolBot is an electronic device that "tricks" a window air conditioner into functioning at cooler temperatures and, according to the manufacturer, with much more optimal energy usage. At this writing, the price of a CoolBot is about $300. Considering that you can use it on successive air conditioners (should the first unit you buy not work out or wear out), it seems like a reasonable investment. The CoolBot website advises sizing the AC larger than the cubic square feet of the space demands in order to reduce load on the motor when cooling. It is important to note that it is always better to buy any cooling unit "too large" rather than undersized, as motors will give out much more quickly when working too hard.

Chilled water lines in masonry wall, Consider Bardwell Farm, Vermont.

With any cooling unit, it is wise to have a backup, or at least a backup plan—you should know if your unit can be replaced or serviced quickly. Consider the possibility of the unit failing at the worst possible moment—what would that cost you in terms of product? Can a replacement be shipped quickly enough to prevent inventory loss? Be sure to size units correctly: If they are undersized and have to work harder than they were designed to, you could be voiding the warranty while you also shorten the system's life span—costing you more in the long run.

What about heating the room in extreme cold climates? If you live in an area of extreme cold in the winter, you may need to provide a source of heat for your aging room. In some cases this can simply be an incandescent light bulb, since even a single 100-watt bulb will generate enough heat to warm a small space. Unless the space is extremely well ventilated, avoid any propane or kerosene heaters, as the fumes will not be healthy for you or your cheese. Small electric heaters with thermostats are another option, although most will not have a setting as low as you need for cheese aging; however, you can plug the heater into a timer and regulate the temperature that way.

Humidifiers

The most efficient way to humidify the air in your aging space will be passively—in other words, with a moisture source that is not mechanical—as any mechanical means will also generate a certain amount of heat. That being said, you may have to choose a mechanical source, should other efforts not provide the relative humidity needed for your cheese.

The most basic passive source of moisture in the air of an aging room is the simple dousing of the floor with a bucket of water. In some cases this will be enough, but in others, maybe not, depending upon the absorption ability of the floor surface and the airflow and temperature in the space. Other effective passive means include terra cotta pots filled with water, or moistened lava (pumice) rocks. You can also plumb a perforated water pipe into the room and drape a sheet of fabric over it—the water will slowly saturate the fabric, and it can then evaporate into the room.

When selecting a humidifying machine, choose cool-fog types—the most popular and effective are labeled "ultrasonic." Instead of spitting heavy water droplets into the air (potentially causing problems on cheese surfaces), ultrasonic units dispense the water in a cool fog that does not cause excess droplets to collect near the humidifying unit. If your water is heavy in minerals, these will be dispersed in the room and form a mineral dust on surfaces. By the same token, be sure that the water supply that feeds the unit is clean and pure, as bacteria and contaminants in the water will be dispersed into the aging room. When buying a humidifier, you should be able to obtain a technical data sheet (either on the Internet or through the manufacturer) that will tell you how many watts the unit will use. If you think in terms of a light bulb, you should get an idea of both the amount of power the unit will use as well as the potential heat it will create.

Air Exchange

In small settings adequate air exchange is often accomplished simply by opening and closing the aging room door throughout the day. If you are willing to monitor your cheese's changing needs, this can work satisfactorily. A more regulated method is to provide an exhaust fan (located near the ceiling so that warmer air will exhaust first) on a timer and a screened and louvered (with flaps that remain closed when the exhaust fan is not on) clean air intake vent. A mechanical timer can be set to provide air exchange as needed. Keep in mind that ammonia fumes will accumulate low in the space. If air movement is not adequate to stir these gases so they can exit through the exhaust vent (which, presumably, will be located up toward the ceiling), then they will not be adequately dealt with.

If you recall the principle of conduction, you will remember that hot air rises, creating air movement. This law can be utilized both to help remove heat from the aging space and to draw fresh air into the room. An exhaust chimney or duct should be installed in the room at a high point and an air intake duct at a low point. These ducts should be filtered and screened to prevent the entry of pests

Fresh air intake piping, poured concrete aging cellar, Bonnie Blue Farm, Tennessee.

and dust. The low intake should be set above the level at which water from washdown of the floor could enter the ducting.

What size should the ductwork be? Most cheesemakers find that it is better to have a pipe that is too large rather than too small; 6 to 12 inches is usually adequate, depending upon the cubic feet of your aging room. Anything under this size will not likely provide enough draw to remove and replace the air in the room.

To further control the airflow, in-line or surface-mounted fans can be used. If installing a fan at either the exhaust pipe or the intake (or both), you can likely get away with smaller-diameter ductwork. Remember to look for wet-environment fans (such as for a shower), as the air in the aging room will be quite humid. Consider wiring the fan's switch to a timer that will allow you to vary the length and frequency of operation. When utilizing an intake fan, you might consider adding an air distribution "sock" to help disperse the air evenly throughout the room. These fabric socks are available from HVAC suppliers and come in different sizes and configurations. Another method to control airflow is the use of an irrigation gate valve that will allow you to restrict the airflow when needed. Some people have mentioned using a home ventilation system to exchange air in aging rooms; however, these units are designed to sense humidity and temperature of both the interior and the exterior air, and they ventilate only when certain parameters are met. For example, one popular brand will not bring in outside air if that air would raise the interior relative humidity above 55 percent—this is designed for the comfort of humans, not cheese. I would imagine that a skilled HVAC technician could bypass some of these parameters, but you might find it simpler to control the air exchange with a timer or manually.

What about the temperature of the fresh air? When you bring outside air into your aging space, it is important to make sure that you do not introduce warm air in the summer and cold air in the winter, unless you are willing to have your cooling or heating unit compensate for this factor (at a higher cost to you). One of the most effective methods for pre-conditioning the air entering the aging room is to run long lengths of ductwork or pipe from the exterior intake underground, at a depth where the ground temperature is stable, before it enters the aging room. If you are building underground, this can easily be accomplished during construction. Piping can be laid alongside the walls, under the slab, or at any other location accessible during construction. Estrella Family Creamery in Washington even ran extra piping through a buried water vault (basically an unused septic tank filled with water) to provide additional air temperature stabilization. If done successfully, this will provide you with year-round fresh air at the proper temperature. If you cannot provide ground cooling, shunt air intake from another room that has as few temperature fluctuations as possible. This may even mean providing two air intake sources, depending upon the seasonal influences experienced by the surrounding spaces.

Shelving

Shelving options in the aging room will depend not only on what you would like to use, but upon what surfaces your inspectors will allow. Across the U.S. people are using, with their inspectors' approval/acceptance, everything from knot-free hardwood or pine (softwood) boards to epoxy-finished or coated metal. Before we focus too much on just what the inspectors will accept, I would like to emphasize that your priority should be what is good for the cheese as well as the cheese-

PVC pipe racking system with plastic bread-bag tags as inventory labels, Pholia Farm, Oregon.

maker. Wood boards may increase the potential quality and romance factor for the marketing of your cheese, but will you be willing and able to keep up with the steps necessary to maintain them as a safe surface for aging a perishable product? If your cheeses will be waxed, vacuum sealed, or plastic (cream wax) coated, then you can probably age them on wooden planks without too much concern from your inspector. (Whether the wood actually adds any tangible quality to plastic-sealed cheese, however, could be debated.)

Plastic or Chrome-Coated Wire
Plastic or chrome shelving is readily available and relatively inexpensive. It will do well for a few years, but in a very humid environment the cut ends and any tears in the plastic coating or abrasions on the chrome will rust. Also, painted metal or chrome-finished wall mounts used with these shelving systems will begin rusting rather quickly, and any wall-mounted hardware that is not stainless steel will also be vulnerable to rust.

Epoxy-Coated Wire
Epoxy-coated wire shelving is more expensive than its plastic-coated cousin but will last much longer. It can be wall mounted or purchased as freestanding shelving. Freestanding shelving of this type is usually assembled in sequence—in other words, the shelves are slid over the vertical supports starting with the bottom shelf and then moving up. This creates a very stable platform for cheeses. It also means that you will not be able to remove any single shelf (except perhaps the top shelf) for individual cleaning. This is not an issue if you have enough space and shelves

Easy-to-clean MetroMax shelving, Pholia Farm, Oregon.

to move cheese to a new set of shelves should you need to do a thorough cleaning. And trust me, if you are making natural-rind cheeses, you will be needing to thoroughly clean those shelves!

MetroMax

My personal favorite shelving is called MetroMax. Plastic, anti-microbial, snap-off grids sit on an epoxy framework. This allows you to remove portions of a shelf for cleaning without having to disassemble the entire unit, or even the whole shelf. They are the most expensive of these three coated-wire options, but in my experience, they are well worth the cost. With all of these shelf types, I recommend freestanding units that can be rearranged as necessary or removed entirely for aging room cleaning.

Wood

What type of wood makes good cheese shelving? While quite a few people are aging cheese on rather porous softwood planks, such as pine and spruce, the best choice is more likely to be a tight-grained, low-resin hardwood. Soft-grain woods are not only more porous—meaning harder to clean and more likely to harbor bacteria—but often more resinous (with pitch or sap), meaning that possible off-flavors may be imparted to the cheese. Some states that allow wood shelves still require that they be coated with a finish, such as polyurethane or wax. This greatly reduces the breathable nature of unfinished wood but still allows for the romance of the phrase "aged on wood" in your product description.

What about cleaning and maintaining wood shelves? A variety of approaches are being used in the U.S. One state requires periodic scrubbing and kiln-heating of the wood to kill bacteria. Vinegar washes (to create an acidic surface not hospitable to most pathogenic bacteria) and/or inoculating the shelves with beneficial bacteria via buttermilk or culture washes are two other techniques used to maintain a good aging surface. Wood shelves should not be scrubbed with a strong chlorine solution, as the chlorine will begin to break down the softer parts of the wood, leaving a more textured, hard-to-clean surface on the shelves.

What about mounting wood shelves? One of the simplest ways to get started with wood shelves is to simply lay the boards across freestanding, coated shelving units. This has the added advantage of allowing the use of thinner pieces of often costly hardwood. When using wood alone, it must be quite thick to prevent warping and sagging under the weight of heavy wheels of cheese. Another simple approach is to build a framework of galvanized or powder-coated metal onto which the boards can be laid. Still another popular method is to install floor-to-ceiling vertical posts or poles into which holes are drilled at different levels. A long, sturdy dowel or pipe is inserted completely through this post, providing a bracket for shelving on each side (see the illustration above). There are many other creative approaches to building wood shelving in aging rooms. Whatever method you choose, keep in mind the following: ease of shelf removal for cleaning; stability of the entire unit when cheeses are placed at various places along the shelf; and rust-resistant hardware and material when building the unit.

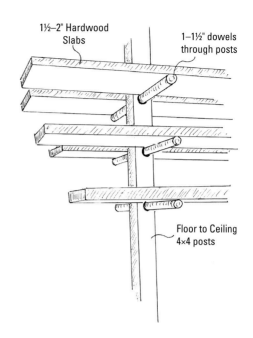

Common wood shelving construction.

Walter Nicolau in his walk-in cooler aging room with wood shelving, Nicolau Farms, California.

PHOLIA FARM

When we wanted to use wood shelving in a state that "doesn't allow" wood, we worked with the state's land grant university dairy department to prove the effectiveness of our methods of maintaining the shelves. We utilized two methods (vinegar and culture washes) on two different types of wood (myrtlewood and Pacific maple) and tracked the shelves' sanitation by routinely sending off swabs for pathogen testing. In the face of evidence, along with the support of academia, it was rather hard for the regulators to refuse our methods. (We waited to implement this plan until we had a couple of years of proven safe product using more conventional aging methods. We felt it was important to set a track record for understanding the process before we "pushed the envelope.")

Wood shelves should not be scrubbed with a strong chlorine solution, as the chlorine will begin to break down the softer parts of the wood, leaving a more textured, hard-to-clean surface on the shelves.

Plastic Matting

You might have already seen the food-surface-approved, plastic mesh matting/netting that cheeses are often placed on while they drain or age. You will also often see this type of matting over the top of plastic-coated/epoxy/plastic shelving as well as wood shelves. This is usually done to increase air circulation around the cheeses. Depending upon the type of cheese and how frequently the wheels are turned, this matting can help equalize the airflow around the cheese. Whether or not cheese sitting on matting over wood actually benefits much from the qualities of the wood (as with plastic-coated or waxed cheese sitting on wood) is a matter of opinion. But the matting could serve as an acceptable surface for the cheese in places where inspectors balk at having it sit directly on wood shelves.

Building Materials and Systems

This is the fun part! When building a new, freestanding aging room, you have a lot of options (depending on what your inspectors will allow) for building materials. Choosing the right ones can be a bit daunting. In the U.S. there are aging rooms built from pre-formed concrete culverts, straw-bale construction, metal sea shipping containers, poured concrete, sprayed concrete ("shotcrete"), refrigerated (reefer) trailers, and walk-in coolers, and of course wood-framed ones as well. Whether your aging room is to be aboveground or buried will greatly influence your choice. Also, keep in mind the "big three" materials issues—cost, longevity, and the qualities the substance might bring (both tangible and intangible) to the aging process—when selecting the material that is best for you.

A charming, old refrigerated trailer provides aging space at Gothberg Farms, Washington.

Reefer Trailers

Let's start with the least romantic aging space, the reefer trailer. A refrigerated truck trailer is a convenient, immediate solution for many creameries. You can find these food-ready units used and ready to go for decent prices. That being said, they are designed to refrigerate to an approximate temperature of 38°F, which is not ideal for most cheeses. They have minimal insulation and a moderate thermal mass. In other words, they will be one of the most energy-thirsty choices you could make for aging cheese. But they serve many fine cheesemakers quite well and can bridge the gap at a time when no other space is available.

Walk-in Coolers

Prefabricated walk-in coolers, with or without refrigeration units, are also easy to find new or used and quick to install. Some manufacturers will help you choose a cooling system for a walk-in that is better suited to cheese. (One company I spoke with was coincidentally asked by another cheesemaker—at the same time—for a quote on customizing a walk-in cooler to cheese-friendly standards. Evidently the salesperson thought we were playing some sort of trick on him and called us "cheese pranksters.")

Florist Coolers

These units are basically like a walk-in cooler, but designed for storing cut flowers. While often cooler than cheese aging temperatures, the equipment is designed to maintain a higher relative humidity than most typical refrigeration equipment. Floral coolers offer limited aging space and so are more often used as an interim solution when no other aging space is available.

Wood Frame

A traditional wood-framed aging room is an affordable, easy-to-build option. If you are designing the aging room to be a part of your dairy building, try to locate it on interior walls; in other words, no wall of the aging room should be a part of any outside wall. This will greatly assist in the control of temperature fluctuations caused by climate. In addition, when pouring the slab of the building, the aging room portion should have a thermal "curtain" (a layer of reflective insulation or other material inserted in the slab between the two spaces) separating the aging room slab from the main slab, to prevent the thermal mass of the slab from influencing the temperature inside the aging room. When building the walls, choose the thickest lumber you can accommodate in the space (and afford). The ceiling should be even better insulated than the walls. As noted earlier in this chapter, the less wood framing, the better the overall R-value will be. The thickness (and R-value) of the insulation you use will be dictated by the thickness of the wall. For example, if your wall is constructed of 2 × 4s you cannot force R-19 insulation into the space, as compacting the fiberglass batts (or alternative material) will negate its insulation potential. I recommend that aging room walls be of no less than 2 × 6 construction, which will allow for R-19 to R-21 insulation. Even better would be to cover the interior of the wall with rigid insulation; the rigid insulation will help mitigate the heat transfer that will occur at each stud in the wall (see illustration).

Ceilings can be easier to "overinsulate" if you have a flat ceiling with space above. A minimum of R-38—which requires 2 × 12 ceiling joists/rafters—is recommended. Again, applying a layer of rigid insulation will greatly reduce the amount of heat trans-

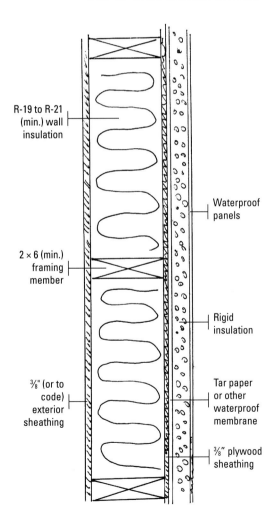

Details for framed wall construction in cheese aging room.

Aging Rooms, Cellars, and Caves

TABLE 10-1: Comparison of Aging Room/Cellar Construction Material Options						
Type	Aboveground or Buried	R-Value (approximate)	Thermal Mass Potential	Initial Cost	Power Usage Average	Story/ "Romance" Value
Reefer Truck	Above	Low	Minimal	Low–moderate	Highest	Low
Walk-in Cooler (No Extra Insulation)	Above	High	Minimal	Low–moderate	High	Low
Wood Framed	Above	High	Minimal	Low–moderate	Low–moderate	Low
Sea Cargo Container	Buried	Depends on depth buried	Minimal	Low–moderate	Low–moderate	Low–moderate
Septic Tank	Buried	Depends on depth buried	High	Moderate	Low if buried to correct depth	Moderate–high
Straw Bale	Above	Very high	Minimal	Low materials, high labor	Low–moderate	High
Pre-cast/ Formed Concrete Culvert	Buried	Depends on depth buried	High	Moderate–high; delivery and placement fees	Low if buried to correct depth	High
Poured Concrete	Buried	Depends on depth buried	High	High	Low if buried to correct depth	High

fer that will occur. Venting the space above your ceiling and insulation will ensure that excess heat will not build up above the space.

With all wood-framed rooms, you will need a waterproofing membrane covering the wood members on the interior. Tar/roofing paper is the usual choice. Even if you are finishing the walls with fiberglass panels (FRPs), or dairy board you should further protect the wood with a waterproof membrane, such as tar paper. If you will be mounting shelves into the wood framing, use a bit of silicone caulking on each screw or bolt to keep moisture from following the hardware into the wood.

Structural Insulated Panels (SIPs)

Structural insulated panels (SIPs) are an excellent choice, when possible. SIPs are prefabricated panels of insulation and structural supports. They have a high R-value and are quick to install. All design needs, such as outlets, window openings, etc., must be determined before the panels are ordered. Finished wall surfaces can sometimes be added at the factory. The panels have a low environmental cost due to the decreased use of lumber, the lack of waste material at the jobsite, and their high insulation factor. If we had it to do all over again, as the saying goes, this is the building material we would have chosen for our creamery and aging room.

Sea Cargo Containers

It is becoming common to hear of using buried sea cargo containers as root cellars. These are large metal containers designed for transporting goods on ships. They are usually not insulated. People have discussed using these for aging cheese, as well. Indeed, at the time of this writing one brave cheesemaker I know has done so—but not with complete satisfaction. Sea cargo containers can be obtained on either coast at a relatively low cost. They are sturdy and can be insulated or even finished on the inside. If you bury one, you will need to reinforce the ceiling or it could buckle under the weight of the earth above. Because these containers were designed for heavy use in salty air, they are extremely durable even when exposed to the elements. Most certainly they are not as long lived as a poured concrete structure, but, given the cost difference, they're possibly worth considering. A metal surface interior is not likely to be adequate for a moist aging room, however, so insulating and finishing would be advisable. Ridged or sprayed-on insulation covered by cement board and plaster would create a surface more closely resembling the properties of a poured cement building or rock cave. Because the containers are long and narrow, I would recommend partitioning off the front section for an antechamber and using the back as the aging room. Building a partition will also help support the ceiling. The entire structure can be draped with heavy-duty pond liner or, even better, liner material designed for use in septic sand filters. This type of construction, however, falls under the *unproven* category.

Septic Tank

Now who would associate the words "septic tank" with an appropriate cheese aging cave? But when you really think about it, what is a septic tank but a pre-engineered, underground room? If you don't need too much aging space, and you have a manufacturer fairly close by, a septic tank can be a viable option. The biggest advantage is that they are designed to be buried; you will have no fears that the ceiling could come down on your cheese—and yourself! To get enough ceiling height, choose a large, single-compartment tank. For example, a 3,000-gallon tank (which is really two 1,500-gallon tanks sandwiched top-to-top) has a ceiling height of about 10 feet and will give you about 600 cubic feet of aging space. A door can be cut after the tank is installed (the tank will come in two pieces), or in some cases the manufacturer might create one during the pouring of the tank at the plant. Septic tanks are already considered waterproof, so you won't need to drape

Septic tank root cellar, Fornaciari home, Oregon.

Aging Rooms, Cellars, and Caves   147

Septic tank root cellar, Fornaciari home, Oregon.

or coat the structure with a waterproofing material. Remember to cut holes for water and power lines, or you can run them through the access hole on top. This method should also be considered *unproven*.

Straw Bales
There are a couple of ways you can use straw bales to construct an aging room. The first is to simply stack bales around the exterior of a wood-framed room or freestanding walk-in cooler as additional, cheap insulation. The bales can be left raw, if protected by the surrounding building, or finished with plaster, as in typical straw-bale construction. It is possible to build an aging room completely of straw bales, but there are a few issues that must be adequately addressed to ensure success. The number-one enemy of straw bales is moisture. The bales must be protected from exterior water by adequate roof eaves. In a residential setting, the walls are usually left "breathable" so water vapor doesn't get trapped in the bales. Unless you are aging cheeses in plastic or wax, your aging room will be a very

> ### RELATIVE HUMIDITY IN STRAW BALES
>
> A large straw-bale winery in California has extensively documented the hygrothermal (moisture and temperature) performance of its earth-plaster-covered straw-bale walls (see www.ecobuildnetwork.org). Part of the winery includes a wine barrel room where relative humidity is often 80 percent (still lower than most cheese aging rooms should be). The winery's study showed that relative humidity within the bales surrounding this more humid room pushed the limit of what is acceptable.

humid environment, so the walls must be sealed from the inside. It is possible to allow the bales to breathe from the exterior or through the top layer, but depending upon your climate and humidity, there could be problems down the road. Moisture sensors should be embedded in the walls at critical levels to monitor bale dryness over time and seasons (this is a common precaution with many straw-bale structures). At the time of writing, I know of no straw-bale aging room more than a few years old; therefore, I would advise that this construction method be approached very thoughtfully (see sidebar).

Concrete Culverts

Pre-cast concrete culverts are rectangular culverts manufactured off-site, then delivered and placed. They can be ordered in many sizes, both in interior height and in length. They arrive on a semi-trailer and are put in place with a crane. They have similar advantages to the septic tank option in that they are engineered for being buried and supporting a great deal of weight, but they are much larger and a better choice for most cheesemaking operations. They are open at both ends, however, so you will still have to construct a back and front wall. Culverts can be placed side by side or intersecting.

Poured/Sprayed Concrete

Poured or sprayed concrete construction will give you the most flexibility regarding design, the longest life span, and possibly the highest initial cost. The expense comes not only from the concrete itself, but from the amount of rebar and steel needed to make it structurally sound. You may be required to have an engineer provide details as to the construction (commonly called "having the plan engineered"), which will also add cost. The weight and composition of earth fill around the cave will dictate wall thickness and thus the amount and placement of steel (rebar) in the walls.

While you may be tempted not to waterproof the exterior of a concrete aging cellar, remember that moisture can be almost as detrimental to unfinished concrete as it can be to wood. Exposure to moisture will cause the inside walls to develop *efflorescence*—moisture moving through the walls carries minerals that are deposited, leaving a white residue. In addition, while the aging room should be buried deep enough to resist freezing, should water enter a crack and freeze, then over time the whole building could be structurally compromised. The interior should be finished smooth and be cleanable. Ideally, you will be allowed to leave the interior unsealed to promote the establishment of an environment that promotes the beneficial microflora and microbial growth desired for cheese

A beautifully designed front entrance for the underground aging cave at Orb Weaver Farm in Vermont.

Poured concrete aging cave ("Cave Beulah") with vaulted ceiling, Estrella Family Creamery, Washington.

> ### FIAS CO FARM'S WHITEWASH RECIPE
>
> Mix in a large bucket (a 5-gallon paint bucket is ideal):
>
> - 3 large coffee cans of hydrated lime (about 12 cups)
> - 1 pound or 1 small coffee can of salt (about 4 cups)
> - 2 gallons of water
>
> To mix this together, mix a little lime and salt, then a little water, then a little lime and salt, etc. If you just dump it all together it's like stirring with a boat anchor. You should let the mixture sit overnight, but we usually just use it right away and have had no problems.
>
> The whitewash should be fairly watery; remember it's a wash, not a paint. Give it a stir once in a while as you use it.
>
> To use the whitewash, just get a big brush and slop it on. Don't worry about getting it on your clothes; it washes out very easily. It may seem like it's not covering very well as you paint it on, especially on new pine 2 × 4s, but it will whiten up considerably when it's completely dry, so be patient. www.fiascofarm.com

Construction of concrete masonry unit (CMU) root cellar with vaulted ceiling, home of Pierre Cloutier, Nova Scotia.

aging. Some inspectors will allow the use of a whitewash in place of a sealing paint. This old-fashioned paint is made from hydrated (masonry) lime, table salt, and water.

If your inspector allows you to use this type of finish, you will preserve the breathability of the walls and add some antibacterial properties (as the coating is quite alkaline). It will need to be recoated periodically.

Concrete Masonry Units (CMUs)

More commonly known as concrete blocks, concrete masonry units (CMUs) are another option for building an underground aging room. CMU construction has some advantages because it is easier for an owner-builder to do and it can be done in stages. CMU walls are not as strong as poured concrete walls, however; they are more likely to have air pockets—open cells that do not get filled with concrete when the walls are filled. Engineering might be required when the wall will be retaining pressure and mass from surrounding geographical features.

Dealing with Unwanted "Visitors"

No, these are not members of the press or public sneaking into your cellar for a quick peek at the cheese; these are creepy-crawlies and other unwanted varmints that affineurs deal with but rarely discuss. While there are certain pests, such as cheese mites and cheese skipper flies, that actually seek out aging cheese, there are other critters that will just be attracted to the warm, moist environment of the aging room. A certain amount of vigilance and diligence is required to keep these unwanted visitors under control.

Cheese Mites

How do you know if you have cheese mites? The first sign of cheese mites will be a fine gray or brownish dust that you will notice on the cheese surface, the floor of the aging room, and shelves. This "dust" is made up of dead and living mites as well as their excrement (yes, mite poop). A quick way to find out if your cheese "dust" is mites is to place a small amount in a pile on a white piece of paper. If you come back in an hour and the dust has spread out, you know it has legs. Given time with your cheese, the mites will begin to pit and crater the surface. Eventually they become more than an aesthetic problem—they actually change the flavor of the entire wheel, giving it a sweetish, herbal flavor. You have probably heard that cheese mites are intentionally allowed to live and work their "magic" on some types of European cheeses, such as Mimolette.

If you come back in an hour and the "dust"

has spread out, you know it has legs.

How do mites get into your aging room? Cheese mites also enjoy feasting upon flour and grains. They can enter an adjoining or nearby building, either by being transported on a host (clothing, hair/body, flies) or by drifting through cracks . . . and eventually find your cheese. If you make aged, naturally rinded cheeses and age them for long enough, you will most likely find mites in your aging room—after they find you!

How do you deal with cheese mites? While some say that mites appear only in cellars that are too dry or too warm, in fact even refrigeration for several days does not kill them. (Don't even ask how I know that. . .) In the past, fumigation with sulfur was used to control cheese mites, but that method has luckily fallen out of favor. Oiling or vinegar-washing of the rind will prevent or slow the progress of cheese mites. Deciding to sell cheese a bit younger has its advantages as well, since cheese mites are slow to affect cheeses under six months of age. Also, keeping the oldest cheeses at a lower level on the shelves will help prevent the mites from "dusting" the younger cheeses. I have also been told that ozone machines (air purifiers) will sterilize the mites, eventually ridding your room of the population. Such units also kill molds and other microflora, however, so this would not necessarily be a good choice when you trying to make a naturally rinded cheese.

Are cheese mites dangerous to humans? Usually not, but there are some reports of dermatitis being caused by contact with the mites. Ingesting them (again, don't ask!) does not seem to be a problem.

Cheese Skipper Flies

How do you know if you have cheese skipper flies? Cheese skippers are named for the legless larvae's ability to contract their bodies and launch themselves several inches into the air, essentially appearing to "skip." The adult females (small, shiny black two-winged flies) lay their eggs in aged cheese and cured meats (quite the gourmands!). The maggots feast on the cheese, causing rapid decomposition. The larvae are fairly large, about 1/5 of an inch long. Between their size and jumping ability, and the state of the cheese when infested, it should be fairly apparent if you have a problem with cheese skippers.

How do you deal with cheese skippers? As the adults are flying insects, you can prevent their entry into the room by utilizing screens on air intakes and antechambers or double-entry doors. Hanging non-insecticidal fly strips or black-light flytraps is also recommended. Thorough, periodic cleaning of the aging room—including shelves and any nooks and indentations where puparia (after the larva has feasted, it will go into a pupal stage before hatching out as a new adult) could hide—will help prevent infestation.

Are cheese skippers dangerous to humans? Ingestion of skipper fly larvae is a cause of severe intestinal problems for humans. The larvae can live in the intestine, dining on the host's tissue and attempting to bore through the intestinal wall. Not a good thing! It's hard to believe that some people intentionally eat a black market cheese in Italy called *Casu Marzu* whose tangy, aromatic, and creamy texture is caused by cheese fly larvae feeding (with the encouragement of the cheesemaker) on the paste of the cheese.

Rodents and Other Pests

Pest control consists of commonsense measures such as tight construction, well-fitted doors and other openings, screening, and filters. If a mouse or other four-legged small pest should get into your aging room, non-poisonous trapping methods should be used, followed by a thorough cleaning and inspection of cheeses for nibbling. For flying insects you can hang non-insecticidal strips and black-light traps. Always keep in mind that not only is the aging room a cozy, moist environment, but it is also well stocked with nourishment for a myriad of unwanted visitors.

You can see that building a successful aging room is a matter of understanding both your cheeses' needs as well as what your geography, climate, and building can offer. This is a complicated topic that is not as well researched and documented in the U.S. as it is in countries where there is a longer history of cheesemaking. I imagine it will be some time before regulatory agencies, industry, and academia here come into alignment on the topic. Until then, many cheesemakers will continue to forge ahead and pave the way—through both their mishaps and their successes.

· 11 ·

Accessory Rooms

There are rooms and areas that can be included in your creamery that will make your work and life easier. In this chapter we will go over some of these rooms and spaces, as well as some of the options you might want to consider when designing them into your building.

Office

Keeping up with the business side of farmstead cheesemaking takes far more time than I ever would have imagined. Invoicing, responding to email, bookkeeping, filing records, and paying bills will take up a significant amount of time in your regular operations. These tasks are unavoidable if you want your business to be a success.

Many people have offices inside of their homes, but creating office space adjacent to the make room will allow the cheesemaker to optimize his or her time and stay a bit more organized than if the office were a part of the home space. An office can also serve as a place to hang lab coats and change shoes before entering the make room; thus it can serve as a "buffer zone" to help keep dirt and contaminants out of the make room.

Consider including a window from the office to the adjoining make room for communicating with family members and other people who might need to talk to you during cheesemaking. If the office is not attached to the make room, think about including an intercom.

Packaging Room

This is one room that we wish we had included in our design. As it is now, we store shipping boxes and materials in the finished loft of our building. When it is time to ship, we bring the boxes and materials to the office. I box the cheeses in the make room and then take them back into the office for weighing and to print labels. If you are not shipping cheeses, then a packaging and wrapping area is adequate. But for storage of cardboard boxes and packing materials, a separate area or room is needed. Cardboard boxes should not be stored in the make room, as they will absorb humidity and begin to mildew. Besides the boxes being ruined by the humidity, they also become nice little environments for pests to lay eggs in or harvest nest-building materials from.

If you are producing soft, fresh cheeses for direct sale to the customer (with little or no shipping involved), then a dedicated countertop in your make room can be used to package and label cheeses for sale. Boxes and carriers can be brought in from the office or a storage area. If you are storing boxes in another room, be sure that room is kept as free of pet hair, dust, etc., as possible and check boxes for debris before bringing them into the make room.

A good separate packaging room will include the following:

- Stainless steel or other food-grade work surface
- Wrapping/packaging paper, containers, wrappers, etc.
- Interior fill material (such as packaging peanuts—recycled or biodegradable, if possible—or paper)
- Labels for product
- "Perishable" labels
- Boxes for shipping
- Postal scale
- Product scale*
- Order book and inventory lists
- Refrigerator for pre-chilling cheeses before shipping
- Frozen gel packs or ice "blankets"

Farm Store/Tasting Room

With the increasing popularity of artisan cheese, I encourage new cheesemakers to think more like a winery or artisan bakery and create a space for visitors to taste and buy your handmade cheeses, learn more about your farm, and perhaps buy other products as well. People are curious about the life of farmers; taking advan-

---

*Rules governing the weighing and selling of products are strictly enforced in most jurisdictions. You will need to check with your state's department of agriculture or other regulatory agency to determine the acceptable use of scales and weights for your application. In general, if a product is ordered by the customer and then weighed and priced by that weight, then the scale must be a "legal-for-trade" scale that is licensed with the appropriate agency and recalibrated at certain specified intervals. If products are prepackaged and sold by the piece, with customers selecting the piece they want to purchase (for example, a small piece of cheese might be labeled as "small—approximately ¼ pound" and there would be a set price for that size), then you might be able to avoid the use of a legal-for-trade scale. If using one allows you to more accurately price your product, however—thereby not giving away an ounce here and an ounce there—then you should consider selling by weight. Again, I really want to emphasize that the level of regulation and enforcement of on-farm and farmers' market sales varies greatly across the country. Be sure to know the regulations for your jurisdiction!

tage of that can help promote both your products and the farmstead cheese sector as a whole.

> **Some cheesemakers have self-service coolers or refrigerators so customers can pick up cheese at their convenience using the honor system.**

A farm store can be as simple as a room with a counter or bar for tasting cheeses (think indoor farmers' market stall) or as ambitious as a fully stocked shop selling local breads, preserves, T-shirts with your farm logo, etc. It can be open only on occasion or regularly. Some cheesemakers have self-service coolers or refrigerators so customers can pick up cheese at their convenience using the honor system. Those using this method find it very satisfactory: it reduces labor related to selling cheese and increases sales by having products available at the convenience of

HOW TO PACKAGE AND SHIP CHEESE

Shipping cheese is one of those topics that can baffle the new cheesemaker—I know it did me! Basically, there is no one preferred method. The only important issue is that the cheese arrives undamaged by either heat or impact.

To help keep cheeses cool during shipping, pre-wrap and chill for 12 to 24 hours before shipping. Use ice packs or blankets placed on *top* of the cheese (because cool air settles) and fill open spaces in the box with a suitable fill material. Double-boxing will also help maintain coolness. Also, consider not shipping cheeses when the outside temperature is above 90°F (at either the origin or the destination)—remember that packages often sit in hot delivery vehicles for many hours before they arrive at their destination. For especially delicate and fresh cheeses, use insulated shipping boxes, either purchased or made at home by cutting pieces of polystyrene insulation (Dow Chemical's trademark name is Styrofoam) purchased from a hardware or building supply store to line a box. Though not very environmentally friendly, these insulated boxes are a necessary evil unless you want to use overnight shipping, which is very cost prohibitive. When we need to use a polystyrene shipping container, we include return postage (very cheap) and ask the recipient to send the container back; this saves us a couple of dollars on a new container—and also keeps it out of a landfill. Let's hope that one day an equally functional yet readily recyclable material will be available.

When packaging cheeses of different weights and hardnesses, pack heavier, harder cheeses at the bottom of the box, fill voids with fill material, add ice packs or an ice blanket, then separate by a cardboard divider that has flaps oriented toward the top of the box (think of it as a topless box). This will keep the weight of the cheeses beneath from pressing against the more delicate cheeses you will place in the top section. By combining the cheeses into one box you can take advantage of the thermal mass (in this case the cool temperature stored within the cheeses) to help maintain an optimum temperature during shipping.

Fill material can be recycled polystyrene or cellulose "peanuts"; shredded paper (beware of ink from newsprint leaking onto cheeses!); straw, hay, or shavings (yes, people really do use these); or crumpled brown paper or newsprint. Whatever material you use, I encourage you to verify with recipients that they practice a recycling program that matches your own sustainability goals. For a location near you that will accept and recycle polystyrene go to www.earth911.com.

Sweet Home Farm sells their entire production of cheese, about 13,000 pounds per year, at their well-stocked, appealing store.

the shopper. Often self-service stands are located not on the actual farm but at a gate or roadside location. This prevents curious visitors from investigating areas of your farm that might be better toured with a guide—or not at all!

If your farm store is attached to your creamery, consider installing a viewing window into the make room. This gives you the opportunity to share more of your operation (and satisfy inquisitive visitors) without people actually entering the make room. If your farm store will be open only when staff is present, it can easily double as an office (of course you probably don't want visitors shopping unattended in a room where your computer, files, checkbooks, etc., might be kept).

Laundry Room

A dedicated (used only for the creamery) laundry facility is very helpful. Of course the home laundry can be used, but it can be challenging to keep creamery laundry free from lint, hairs, and other debris that is common in household laundry. Cheesecloth can be hand-washed and sanitized in the creamery, but lab coats will need a washing machine. Many cheesemakers are also using small terry cloth towels, instead of paper towels, for drying hands. These meet PMO requirements for single service, as they are used once and then re-laundered. The initial investment is higher than for paper towels, and there are additional expenses for hot water, detergents, bleach, and wastewater disposal. Paper towel costs continue to rise, however, and disposal (unless you are planning to compost them) is an issue. We wash and dry our small hand towels separately from the lab coats and cheese cloths so it's a quick and easy process to fold and put them away. After folding, they are bagged in reusable storage bags and stacked near each hand-washing sink. We started out using single-fold paper towels, the type you pull out individually from a wall-mounted dispenser. I took those dispensers off the wall, turned them upside down, and remounted them. The front opening (that was designed for refilling the unit with paper towels) now serves as a self-closing door that you can flip up to remove one cloth towel at a time.

Toilet

A toilet facility is mandated by the PMO. It should be located convenient to the milking parlor and milkhouse. For many small creameries, a bathroom in the residence is perfectly acceptable. But remember, the inspector will need to have access without your presence for inspection purposes. The toilet facility does not have to be state of the art; it can even be an appropriately dug "outhouse" or portable chemical toilet. It will be inspected for proper hand-washing facilities; general cleanliness; a tightly fitting, self-closing door; and absence of pests.

When designing your creamery, consider building the bathroom to meet the standards of the Americans with Disabilities Act (ADA). Go to www.access-board.gov or talk to an ADA representative or your attorney for information on current dimensions and standards, as well as advice on meeting other requirements, such as parking (if your farm will be open to the public on a regular basis). Organizations that receive federal or state funding may not be allowed to include your farm (for membership, tours, etc.) if these standards are not met. While you may currently have no plans to work with such groups, there is no reason not to include the possibility into your building plans, if you have the chance.

Medicine Storage

Medications, treatments, and supplements for your animals must be stored in a place that the inspector can access. Our medicine cabinet for the goats is in our laundry room, through which the inspector also must pass to check on our toilet facility. The inspector will also look through the entire barn and feed rooms. If you have other species that you do not milk, such as horses, pigs, etc., be sure to keep their medications, de-wormers, and supplements in a specific location that is identified with a label or sign with something like "horse shelf."

The inspector will be looking for properly labeled medications that include a species-appropriate listing (such as for goats, if you milk goats, and not medication that is labeled only for cows), name of medication, milk withholding times, dosages, expiration date, and purpose. If you milk goats or sheep and use medication that is labeled only for cattle, then you will need what is called "extra labeling" from a licensed veterinarian. This is true even if the medication is available at the feed store and is widely used for your species. The extra label will include the above-mentioned facts as well as the veterinarian's name.

If you keep medications that require refrigeration, expect their storage area to be inspected as well.

Keep any medications that are only for use in young or male animals (in other words, non-lactating animals) on a separate shelf or case and label them accordingly.

Your creamery is now complete! Don't be surprised, however, to find things you have left out or would do differently. I don't believe that there is any way to fully anticipate the future and changing needs of any business, much less one that involves both animals and food production. We all have a "if we could do it over" list. I hope this section will have helped you to one day have a shorter list than most of us!

PART IV
LONG-TERM SURVIVAL GUIDE

· 12 ·

Safety: Why "It Hasn't Killed Anyone Yet" Isn't Good Enough!

I have to admit, it took us a few years to get around to even *trying* to create a HACCP (Hazard Analysis and Critical Control Points) plan. It is a daunting task made more difficult by a lack of clear, simplified instructions. Many publications on the subject are designed for the large-scale industrial dairy, not the little creamery operated by only a handful of people. A good portion of the documentation steps in a formal HACCP plan are meant to track the performance of a crew of workers, making these steps seem like wasted time and effort for the very small creamery. Indeed, the plethora of steps and paperwork is very likely to discourage many from attempting to create a solid food safety plan. With that in mind, I have simplified the process for those who would like to get started with HACCP. Should you become interested in a more complex and thorough program, resources listed in appendix A and in the notes/references section will provide much more in-depth discussion of the topic. The subject of product recalls—a part of any good quality assurance plan—is also dealt with in this chapter. While recalls are something we can hope to never experience, the reality is that you or someone you know will probably have to deal with one at some point. The old saying *"An ounce of prevention is worth a pound of cure"* definitely is pertinent when it comes to quality assurance programs!

What Is HACCP?

HACCP (pronounced with a short "a" as in the word "at," with a soft "c": "HA-sip") stands for Hazard Analysis and Critical Control Points. In a nutshell, HACCP is a system designed to define every step of a product's manufacture, identify the points in the process that are critical for food safety, and create a plan that delineates and documents safe manufacturing processes. It is a preventive program, focusing on the process rather than on product testing. An important difference to note: HACCP is not about quality as it relates to aesthetics—taste, texture, aroma, etc.; it is only about the *safety* of the product.

THE SEVEN PRINCIPLES OF HACCP

1. Conduct a hazard analysis
2. Determine critical control points
3. Establish critical limits
4. Establish monitoring procedures
5. Establish corrective actions
6. Establish verification procedures
7. Establish record-keeping and documentation procedures

In the United States, HACCP programs are mandatory in certain food industries, such as seafood and fruit juices. At the time of writing this book, they are not required in the dairy industry, but they are encouraged. Furthermore, larger dairy processors are usually required by their clients to have HACCP plans in place. Even if you do not create an in-depth HACCP plan, you can still use the principles of HACCP to create your own quality assurance program.

HACCP is not about quality as it relates to aesthetics—taste, texture, aroma, etc.; it is only about the *safety* of the product.

Before a HACCP plan is developed, you must first define your "prerequisite programs." What is often confusing is that while prerequisites must be defined as a part of building a HACCP plan, they are not part of the seven official principles of HACCP as defined by the National Advisory Committee on Microbiological Criteria for Foods (NACMCF).

Alyce Birchenough checks and logs the pH during her cheesemaking process as a part of Sweet Home Farm's quality assurance program.

To further add to the bewilderment, there are steps in prerequisite programs—as outlined by the Dairy Practices Council's Guideline No. 100, "Food Safety in Farmstead Cheesemaking"—that overlap steps and principles in HACCP.

To help make all of this easier to understand and to encourage the creation of a quality assurance program that even the smallest cheesemaker can implement, I have simplified and combined prerequisite programs and the seven HACCP principles into three steps.

Three Steps for Creating a Quality Assurance Program

1. **Prerequisite Programs**: These are standardized procedures based on knowledge of proven facts regarding the safe manufacture of a food product. They include *Product and Process Information, Standard Sanitation Operating Procedures (SSOPs),* and *Good Manufacturing Processes (GMPs).*
    a. **Product and Process Information**: Includes a description and a flow chart (showing every step of the process) for each cheese made.
    b. **SSOPs**: List step-by-step proper cleaning and sanitation of equipment and facilities.
    c. **GMPs**: List procedures for maufacturing related to facility, equipment, and processes.
2. **Hazard Analysis**: Each step of your process is analyzed for any chemical, biological, and physical risk associated with that particular step.
3. **Critical Control Points (CCPs)**: These are points during the process when a "control" (think a step, a procedure, etc.) is used to prevent, limit, or eliminate a critical food safety hazard.

To sum it up, the basic principle of HACCP is this: All hazards (identified in Step 2, Hazard Analysis) are controlled, whether through Step 1 (Prerequisite Programs) or Step 3 (Critical Control Points). If you have identified everything that could possibly go wrong and you develop procedures to prevent all of these potential problems, then the odds are very good that you will be producing a safe product.

The foundation for a good HACCP plan is knowledge. Until you are well educated in the microbiology and chemistry of the cheesemaking process, including cleaning and sanitizing, you cannot adequately understand or foresee the potential hazards that exist. This is another reason that even the most talented, intuitive cheesemakers should seek ongoing education about their craft!

Until you are well educated in the microbiology and chemistry of the cheesemaking process, including cleaning and sanitizing, you cannot adequately understand or foresee the potential hazards that exist.

So let's assume you are a fairly experienced cheesemaker with some education regarding milk chemistry and cheesemaking microbiology. In addition, you have books and publications that delineate good cleaning and sanitation, basic good manufacturing processes, and risks and hazards associated with taking raw, living milk and turning it into a fermented food. Where do you start?

Step 1: Prerequisite Programs

Step 1a: Flow Chart and Product Description

Creating a flow chart for each cheese you make (start with just one for now) is the most logical starting point for most cheesemakers. The flow chart falls under the "Product and Process Information" portion of the prerequisite programs of HACCP. When creating the chart, be sure to list every step of the process, any added ingredients (in addition to milk), and the product and by-product distribution.

You can also include a more complete description of the cheese that includes the following:

- **Common name**: Does the cheese fall under a *Federal Standards of Identity** description?
- **Commercial name**: What do you call it?
- **Packaging**: How is it packaged?
- **Labeling**: How is it labeled?
- **Shelf life**: What is the shelf life after sale?
- **Storage temperature**: What is the correct storage temperature?

Here is an example using one of our cheeses, Elk Mountain:

Common name: None
Commercial name: Elk Mountain
Packaging: Sold in whole or cut wheels wrapped in freezer paper and/or plastic wrap
Labeling: Handwritten on paper with name and date of make
Shelf life: 4–6 weeks, depending on size of cut
Storage temperature: Aged 6–9 months at 55°F; storage after aging, 38°F

Step 1b: Standard Sanitation Operating Procedures (SSOPs)

SSOPs deal with the proper cleaning and sanitizing of everything in your dairy and creamery, including workers' hands, processing equipment, floors, refrigerators, and raw material.

**Federal Standards of Identity* for dairy products lists products with production standards that must be met for that cheese or dairy product. If you make a cheese that falls under any of these descriptions, you must meet the production criteria for that product, and it must be labeled accordingly. (For more on the Federal Standards of Identity, see chapter 3.)

Flow chart for Pholia Farm Creamery's Elk Mountain cheese.

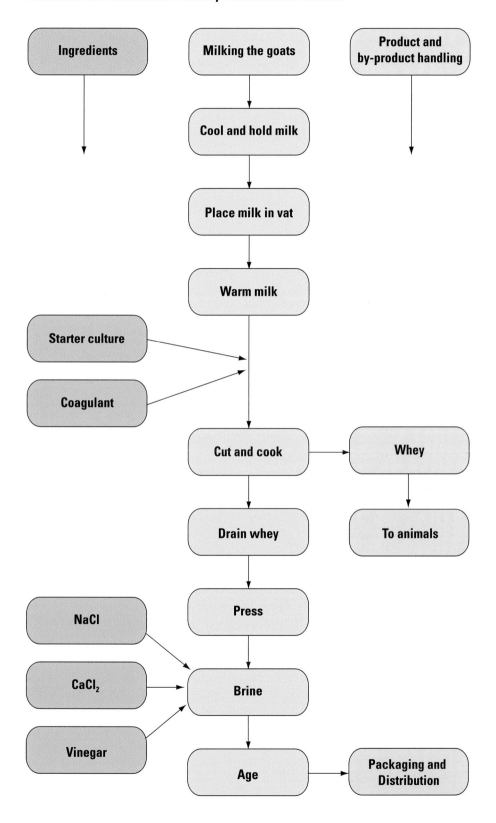

SSOPs—Things to Remember
1. You cannot sanitize something that is not clean; therefore proper cleaning is as important as proper sanitizing.
2. Proper dilution of detergents and sanitizers must be coupled with proper water temperature, scrubbing, and length of time cleaned to be effective.
3. Sanitizers not diluted to the proper concentration will be ineffective if weak and leave residue (a chemical hazard) if too strong.
4. Good SSOPs include documentation of cleaning frequency and descriptions of cleaning protocols and who will do the job.
5. To be effective, SSOPs must be monitored and corrections made—choose a very reliable person (probably yourself) for this job.

SSOPs—The "Big Eight" Sanitation Concerns
1. Safety of water that comes into contact with cheese (such as in washed-curd cheeses) and with cheese contact surfaces (vat, molds, etc.).
2. Condition and cleanliness of cheese contact surfaces, utensils, gloves, etc.
3. Prevention of cross-contamination between finished cheese and raw products, utensils, packaging, etc.
4. Prevention of adulteration of cheese, cheese contact surfaces, utensils, gloves, packaging, etc., with chemical hazards, such as cleaning compounds and lubricants; physical hazards, such as bandages, fingernails, jewelry, and pests; and biological hazards.
5. Maintenance of hand-washing and toilet facilities (this includes single-use towels and hot water).
6. Correct labeling and storage of all chemicals.
7. Control of worker health conditions that could cause contamination of cheese, packing materials, surfaces, etc.
8. Exclusion of pests and pets from dairy and creamery.

Step 1c: Good Manufacturing Practices (GMPs)
GMPs deal with policies regarding the production of cheese as it relates to everything in your dairy and creamery, including hygiene, quality of equipment, construction of floors, quality of refrigerators, and documentation of raw materials. GMPs are a part of the Federal Code of Regulations (see resources in appendix A).

GMPs—Four Main Areas of Food Processing
1. *Personnel hygiene* includes policies on hand-washing, jewelry, eating in manufacturing areas, illness, cleanliness of clothing, etc.

2. *Building and facilities* includes lighting and ventilation, hand-washing stations, pest management, etc.
3. *Equipment and utensils* must meet food-grade standards and be easy to clean, sanitize, and maintain.
4. *Production and process control* includes time and temperature logs, records on ingredients, lot identification of cheeses, etc.

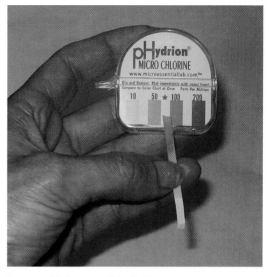

Common chlorine strength test strips.

Comparing SSOPs to GMPs

To better help you understand the difference between an SSOP and a GMP, table 12-1 presents two examples, one using milking equipment and one using starter culture. You can see how the two prerequisite programs overlap in many subject areas, but the focus of each is different. When used together, a complete process that creates a safe product is the result.

Putting the Prerequisites Together

Once you understand the goals and parameters for SSOPs and GMPs, you are ready to document them in your prerequisite program. I have seen this done a couple of ways. Some people write a narrative for each step of the flow chart that

TABLE 12-1A: Pholia Farm Milking Equipment	
SSOP	GMP
Clean and sanitize after each use:	All milking equipment is:
1. Pre-rinse milking lines with 100°F water, minimum of 3 cycles. Hand-rinse bucket, lid, and strainer with same. 2. Drain rinse water. 3. Fill sink with 140–150°F hot water and detergent in correct proportion. Hand-scrub bucket, lid, and strainer with same solution. 4. Run wash cycle 5–10 min. Not to less than 120°F. 5. Drain wash water. 6. Fill sink with 110°F water and acid rinse in correct proportion. Rinse bucket, lid, and strainer with same. 7. Run rinse cycle minimum 3 min. 8. Hang bucket, lid, and strainer to dry. Disconnect milking lines from washer and hang to dry. Inspect for effectiveness of cleaning; re-clean if necessary.	1. Acceptable for Grade A raw milk. 2. Easily cleaned and sanitized.

TABLE 12-1B: Starter Culture	
SSOPs	GMPs
1. Storage area is properly cleaned and sanitized. 2. Storage container is properly cleaned and sanitized.	1. Upon receiving, product is logged, noting lot number and expiration date. 2. Product is stored at proper temperature. 3. Temperature of storage area is logged daily.

describes each process and states what SSOPs and GMPs apply. Others document a full set of SSOPs and GMPs in a binder. During Step 2 of creating a quality assurance program (Hazard Analysis) you can refer either to the narrative that follows the flow chart or to your prerequisite program binder. If you skip the creation of a well-documented prerequisite program, you can still complete the hazard analysis, but it will be less thorough and—even more important, perhaps—less impressive and persuasive should you ever find an FDA official at your doorstep.

Step 2: Hazard Analysis

Once you have your flow chart and are comfortable with the standards for SSOPs and GMPs (keep reference books and publications on hand to help with the process) you will be ready to analyze each step of your process for hazards in three categories:

- **Chemical**—examples include antibiotics, sanitizers, pesticides, paint, and allergens.
- **Biological**—examples include pathogenic bacteria, such as *Listeria* and *E. coli*; viruses, such as hepatitis A; and mycotoxins from mold.
- **Physical**—examples include bandages, fingernails, disposable glove fragments, glass, and pests.

The Dairy Practices Council's Guideline No. 100, "Food Safety in Farmstead Cheesemaking," contains very thorough lists of potential hazards in the above categories. In addition, it covers SSOPs and GMPs as they relate to reducing and/or eliminating the risks posed by these hazards. It is an inexpensive publication well worth having.

So let's take a couple of steps from the flow chart for Elk Mountain cheese and analyze them for risks.

Remember, when you are working on control measures for each step, measures must deal with safety only—not quality as it refers to taste, texture, aroma, etc. *A good HACCP plan does not mean a tasty cheese, only a safe one!*

Step 3: Identifying Critical Control Points (CCPs)

This part of the HACCP process always confused me. I wondered why you would need to identify additional hazardous points if you had created a sound prerequisite program. If your SSOPs and GMPs dealt with all of the hazards you could identify, why would you need to label certain areas with a CCP? It didn't make sense to me, until I understood the difference between control measures (also called control points) and critical control points:

- **Control measures** are points in the system where loss of control may result in a reduction of product quality, but not a reduction in the safety of the cheese.

TABLE 12-2: Hazard Analysis for "Milking the Goats" Step, Pholia Farm Elk Mountain Cheese

	Potential Hazards	Control Measures
Chemical	Sanitizer, detergent, and acid rinse residue on milking equipment; antibiotics, medications	Proper dilution of sanitizer, proper rinsing Records of does on medications, milk withholding time noted in parlor, doe prominently identified
Biological	Pathogenic bacteria contamination from dirty teats, first milk, subclinical mastitis, improperly cleaned milking equipment, hands of milker	Proper cleaning, sanitizing, and storage of milking equipment; proper pre-milking, teat cleaning Hand-washing for milker, gloves during milking; food-grade, easily cleaned and sanitized milking equipment; routine somatic cell count (SCC) on all does, perform California Mastitits Test (CMT) on suspects, records on does with milk below standard, proper disposal of milk
Physical	Hairs from goats, milker; flies	Proper cleaning of teats Disposable gloves on milkers, filtering of milk with one-time-use filters

TABLE 12-3: Hazard Analysis for "Brine" Step, Pholia Farm Elk Mountain Cheese

	Potential Hazards	Control Measures
Chemical	Cleaner and sanitizer residue in brine tank after new brine is mixed	Proper cleaning, rinsing, and sanitizing with correctly diluted sanitizer
Biological	Inadequate pH development before brining, contamination when moving cheese from drain table to brine, improperly maintained brine	Cheese pH checked prior to brining, minimum pH of 5.4 or lower, logged in make sheet, hand washing before handling cheese, brine fully saturated
Physical	Adulteration when moving cheese from drain table to brine	Hair net and clean lab coat when in cheese room, personal hygiene policy in place, pest management policy in place

- **Critical control points** are points in the system where a control must be exerted or food safety will be compromised. CCPs represent a point in the process where no other measures that will occur later can eliminate or mitigate the hazard.

So think of CCPs as a kind of red flag or highlighted control measure. Examples of CCPs in the cheesemaking process are pasteurization, acidification, salting, and aging. Once you have done your hazard analysis, you can go back through the points and mark as CCPs those that present a critical risk. Once you have determined where the CCPs are in your process, it is important to set parameters and document that they have been met. This is the whole point of HACCP.

Is a Recall in Your Future?

A cheesemaker I once interviewed (who is no longer in business) gave me a provocative quote: "If you stay in business long enough, you will have a recall." Stories abound regarding recall nightmares, some with happy outcomes and others from

> ## FDA RECALL CLASSIFICATIONS
>
> *Class I:* There is a strong likelihood that use of, or exposure to, a product will cause serious adverse health consequences or death. (For example, contamination with pathogenic bacteria or Class I allergens not declared on a label.)
>
> *Class II:* Exposure to, or use of, a product may cause temporary or medically reversible adverse health or consequences or where the probability of serious adverse health consequences is remote. (For example, contamination with a piece of glass.)
>
> *Class III:* Exposure to, or use of, a product is not likely to cause adverse health consequences. (For example, a product that contains an ingredient that is safe but is not declared on the label.)

which companies could never recover. Even the ones with a happy ending were stressful and costly while they were happening. If you have developed a good HACCP or other quality assurance plan, you will have a very good chance of surviving an FDA-requested recall; if you do not, then the odds are not as good. In addition to the HACCP plan, you should have a recall plan—indeed, it is considered a part of a complete prerequisite program.

Recalls Defined

Be wary of the use of the term "recall." It should be used only when a product poses a risk hazard as defined by the FDA guidelines (see sidebar). Use it only when an FDA-requested, or FDA-instigated, recall is taking place. Products can be brought back from customers under a "market withdrawal" plan. When you read the classifications in the sidebar, please pay close attention to Class III. Even at this lowest level of FDA recall your business could be at risk for something so seemingly simple as putting the wrong label on a cheese!

> ## RECALL STEPS
>
> 1. **Contact lawyer and FDA:** If you discover cause for a true recall (your situation meets the criteria listed in the sidebar on FDA recalls), contact your lawyer and the FDA.
> 2. **Contact** retailers, wholesalers, and distributors and arrange to have your products returned. Follow this initial contact with a formal recall notice (see step 5).
> 3. **Obtain samples** and send them to an independent lab for testing (if the FDA instigated the recall).
> 4. **Public notification (press):** *only* if the product poses a serious health threat, i.e., Class I.
> 5. **Recall notice:** Create a notice that identifies the product (lot number or other identifier), the reason for recall, who to contact at your company, what to do with the recalled product, and reimbursement procedures.
> 6. **Terminate** the recall after all product has been accounted for and after consulting with the FDA.

What Can You Do to Prepare Right Now?
There are several fronts on which you can prepare for the potential of a recall. All of these steps are sound business measures—even without fear of a recall—so implementing them, regardless of how safe you feel, is still a great idea.

Preventive Measures
1. **HACCP:** Nothing will help set the stage for your defense better than a sound HACCP plan!
2. **Product identification** in-house and tracking: Identify each batch of cheese on your make sheet/log and track it through its aging, packaging, and sale. This step is essential in order to perform a successful recall or market withdrawal.
3. **Shipping records:** This follows product identification.
4. **Company contacts:** Know who to contact for each retailer, wholesaler, or distributor that sells your cheese.

Drills
Hold a mock recall, go through all the steps below, and see if you are confident that each step can be completed with relative ease. Remember, if a recall situation arises, the stress level will be high and responsibilities that would seem simple under normal conditions become challenging.

There, you've made it through a tough chapter on topics that are all about the harsh realities of running a business that produces what is thought of as a high-risk product. Even if you are not ready to implement a quality assurance and recall plan right now, keep it in mind and take the time—even if you do it in stages—to create what I like to call your "business armor." You have already put a lot of thought, time, and, no doubt, money into this venture; you might as well take the extra steps to protect it for the long haul.

· 13 ·

Increasing Your Bottom Line: Classes, Agri-tourism, Additional Products

At some point in your development as a business, you may find that you want to augment your income, either to meet increasing costs or to make a bit more money—or even as a way to vary your work. While the usual route would be to increase production, this is not always the best choice. You may feel that your land and time do not allow for an increased herd size without your having to compromise quality of animal care. Or perhaps you know that increasing the number of animals you milk will necessitate hiring outside labor, and this is something you don't want to deal with. For many of us it is easy to lose sight of our original goals when our bank account is screaming "Too low!" If growth is not your path for sustainability, then a value-added option might be the right choice. (Technically, cheese is already a value-added product to the commodity product of milk, but as this book is dedicated to those who wish cheese to be their primary creation, I will not refer to it as "value-added.")

Keep in mind that the ideas in this chapter are just that, ideas—not complete guides to teaching classes or running a farm stay. If you find any of these ideas intriguing, you can explore them through some of the recommended resources in appendix A or through your own investigation and experiences.

However you approach value-added options, be sure to start with regulatory issues first. For each of the following examples there will be steps that you must take to ensure that your enterprise is functioning legally. You will also need to address liability issues by discussing the venture with your insurance agent. A good idea can be great only if you can conduct it legally!

Agri-education: Classes and Mentoring

One of the simplest ways to bring in a little extra cash is to teach cheesemaking, herdsmanship, or farming classes, either on your farm or at events, such as outdoor festivals or fairs. If you enjoy sharing your experience and knowledge, not only can teaching bring in a little extra cash—it can also be very gratifying.

Classes given at your farm will offer the largest profit margin, as well as being the least difficult to support logistically, since your equipment and supplies will

BLACK MESA RANCH

Black Mesa Ranch (BMR) is a 280-acre off-grid Nubian goat cheese dairy owned and run by Kathryn and David Heininger, located off the beaten path near Snowflake, Arizona. David and Kathryn's approach epitomizes, in my opinion, the perfect combination of resources, skills, and passions. They have turned their love of goats, cheese, chocolates, and a remote lifestyle into a viable business. David explains how agri-tourism has worked for them:

"Holding workshops allows us a way to channel the requests for information and education we get into a planned, organized, and billable format that amounts to about 8 percent of our income. We host only about five or six three-day workshops a year. It is just too disruptive to our full-time dairy business to try to do more. The workshops are full-immersion, intense sessions that leave most of our guests taking advantage of our optional afternoon breaks. Most are attended by couples, but we have also hosted a few families and groups of chefs. It has been gratifying to see what some of our guests have done with their knowledge—including opening new creameries with their BMR workshop graduation certificates proudly displayed!

"We feel a big responsibility to provide our workshop guests with good value, so there is definitely a level of concern that something unforeseen might go wrong during their stay that would negatively affect their experience. Equipment failure, a sick animal, or other problems could prevent us from giving the guests our full attention. On the other hand, these are also real-life parts of the business of which the guests need to be aware if they are going to be successful in their own ventures.

"One common misconception that people who are considering adding workshop-type activities to their operations seem to believe is that by having other people (guests) doing some of the work, it will lighten their own workload. It will, in fact, be quite the opposite, but the entire experience will still be well worthwhile for all concerned."

---

not have to be transported to another location. Classes can be tailored to your own level of expertise and interest. Depending upon the climate where you live, try to schedule classes for a time of year when people are not engaged with a lot of outdoor activities. We have found that late fall and winter is the time when people are looking for something to do indoors. This also may be the time of year when you are not as busy with the farm and when, if your cheese market is seasonal, a little additional income could be helpful.

Teaching off-site is also an option; in fact, if you are involved with any groups or clubs, you can almost count on being asked at some point to teach a cheesemaking class. Determine ahead of time what you will need to provide (such as milk and equipment), the conveniences available in the room (slide presentation equipment, sinks for washing, hot plates, etc.), whether the group (or you) will pay for handouts you prepare, and what kind of stipend you will receive. Don't be shy about clarifying all of these points before accepting—it is the professional approach and should be respected and appreciated.

If there is a nearby university or college, you might also consider partnering with it to provide educational and personal enrichment experiences. Often summer programs through colleges offer classes related to local agriculture for youth, seniors, and others. While these types of classes will not usually provide the same level of income as when you administer them yourself, they can be a valuable community-building tool.

David Heininger (left) teaching a cheesemaking class at Black Mesa Ranch, Arizona.

At Pholia Farm we offer animal husbandry and cheesemaking mentoring as a way of providing education for those wanting one-on-one hands-on experience with livestock and/or cheesemaking. By offering animal advice for a fee, we are able to help those who are both serious about learning and understanding of the limitations on our time. As a way to help those who are not able to pay and/or visit our farm, we try to provide as much information as possible on our website regarding our herd management and farm. If you should consider mentoring/tutoring, base the price you charge on both your time as well as the opportunities you will give the student. If you prefer a less intense, activity-packed day, then charge accordingly. The experience should be worth both your time and the student's money.

Agri-tainment/Agri-tourism: Tours, Events, and Farm Stays

Terms like "agri-tainment" might make some of us who take our farming very seriously cringe a bit, but there is nothing disreputable about taking advantage of the opportunity provided by people who are seeking amusement and entertainment in a farm setting. The University of California at Davis (UC-Davis) offers many helpful tips on agri-tourism at www.sfc.ucdavis.edu/agritourism, including advice on tours, events, and farm stays. Your own state department of agriculture or agri-business may also offer helpful tips and guidelines.

Tours

Unless you lock your gate and remove your phone number from anything that relates to your farm, school groups, food writers, foodies (see chapter 1 for more on foodies), and urban dwellers seeking a farm experience will be calling you to ask about tours. There are some ways to deal with this ahead of time that will make your life easier and perhaps satisfy the curious. Here are some of the issues you will need to decide upon:

- *Will we charge a fee?*
- *Will we provide cheese tasting?*
- *Will we sell our cheese? What about other products?*
- *Will we offer tours on a schedule or by appointment?*

Fees

While most farms do not charge a fee for a quick tour, you may want to consider a small fee if you are providing an educational component and tasting, especially for groups. We have found that individuals who visit us usually purchase enough cheese and other products to make it worth our time, while very few groups do. Groups, such as seniors and school groups, are usually there for entertainment and educational purposes. There is nothing wrong with coming up with a small fee that you feel will compensate you for your time—in fact, you will probably find yourself providing a better tour and feeling better about it if you develop a fee schedule. We provide a "tour tip jar" instead of a fee and casually let people know that it is available if they would like to leave anything. These tips go to our daughter, who is also the tour guide.

Tastings

For the full experience, your guests will probably want to try your cheese. If you have included a tasting room/farm store (for more on farm stores, see chapter 11), you will be prepared for such events. If not, be sure to provide protection for the food from flies and dust and have hand sanitizer or other hand-washing facilities for your guests. If you do not have cheese for sale at your farm, consider a fee for tasting. Remember that wineries often provide one free tasting and then charge for additional samples. While I do not want cheesemakers to completely emulate wineries, I do think we need to learn to value our time and products at least as much as the winemakers value theirs!

Products

As a farmer, you should have the right to sell products legally produced on your farm, and you may be able to sell other products as well. T-shirts and mementos for visitors to commemorate their time on the farm are always popular and can be a way to add a little income as well as increase your public profile. If you are selling products produced by others, you will want to find out if you need a business license.

Scheduled Tours, or By Appointment Only?

There are advantages and disadvantages to being open on regularly scheduled days versus by appointment only. If you are off the beaten path, being open on regular days can be a study in "hurry up and wait." It's not as bad if you can continue to work while waiting for customers, but if you have to stop everything and sit behind a counter—and no one comes—then it can be a waste of valuable time. On the other hand, if you have even just one regular time that you are open and do not want to deal with tours by appointment, then it is very handy to simply tell people that they will have to visit during that time. If you decide to be open "by appointment only," be ready to say no to people when it is not a good time for visitors—this can be harder than you might think! Very pleasant, tough-to-refuse people will call who are "only in town today" or "just passing through" or who have "a food-related blog" or some other persuasive angle that makes it difficult to turn down their request. Visits by appointment can sometimes mean setting aside time and waiting for people who arrive late, leave late, or sometimes don't even show up.

Goat milking contest at Pholia Farm open house.

Baby goat feeding at Pholia Farm open house.

Events

Farm events can include open houses, a harvest festival, a corn maze, or a pick-your-own-pumpkin patch. You can also develop your own event that takes advantage of another community happening, when visitor traffic and press exposure might be at its optimal.

Open House

Holding an annual open house can be a great way to offer a peek inside your life on one well-planned, very busy day. While you may not reap immediate financial rewards from such an event, you will increase your community profile and relationship. I find it is also a great way to get the place spiffed up!

Here are some tips for putting on a really great open house:

1. Choose a time of year when the weather is pleasant.
2. Have baby animals to see and pet.
3. Invite a local winery, brewery, etc., to offer samples of its products.
4. Have lots of helpers for keeping track of visitors and answering questions.
5. Have hand-sanitizing stations near animal pens.
6. Securely latch and "visitor-proof" all areas where risks might exist (including animal pens).
7. Hold several mini-events throughout the day:
    a. Milking contest (provide blue ribbons with your farm name to all participants)
    b. Baby feeding time
    c. Name-the-baby contest
    d. Hay rides
    e. Story time for kids in the hay barn
8. Have at least one person available to help with parking.
9. Provide water and snacks for your helpers and volunteers.
10. Send out a good press release four weeks before and again one week before the date of the event. Include photos from last year's event.

Photo opportunity at Pholia Farm: Author's daughter Phoebe Caldwell and nephew Jacob Simmie as "goat and milkmaid."

Farm Stays

Farm stays have been popular in Europe since the 1990s and are catching on in the U.S. A less expensive version, the "hay hotel," is becoming popular in Germany (people bunk down on bales of hay in communal rooms on a working farm) but probably won't be seen in the States for some time.

Farm stays differ from bed-and-breakfasts mostly in their emphasis on having a rural experience on a working farm. Some jurisdictions also more severely limit the number of rooms and guests that a farm stay can have, in comparison to a B&B. Meals are sometimes included, and in other instances guests are provided with a stocked kitchen. Usually a commercial kitchen is not required. While the

A lovely, rustic, and comfortably appointed cabin draws tourists to stay at Bonnie Blue Farm in Tennessee.

farm stay is a newer concept here is the U.S., many areas are encouraging them as a way to help farms and farmers remain viable.

Be sure to check with your county or local jurisdiction to find out what rules might apply should you consider offering farm stays. There are quite a few resources for getting started with a B&B, and many people who offer farm stays refer to these guidelines; this is a place to start, but be aware that regulations will likely vary for agricultural-based tourism. In some states, if you have only one or two rooms you may not need a license, but you will still have to comply with the conditions required for licensing. Some states will require you to collect a lodging or transient occupancy tax (TOT), the same as is collected by hotels, campgrounds, etc. A good place to start asking questions is your local agricultural extension agency office, the agribusiness council/department for your state, or even your state department of agriculture. While they may defer to local jurisdictions, they often have helpful websites, publications, and links designed to support agri-tourism in your state.

If you are considering farm dinners, start making notes as to what items are available seasonally in your area— this will help you optimize the timing of the events.

Farm Dinners

Another growing agri-tourism activity is the farm dinner. For foodies, the idea of eating outdoors (or in a barn or such) on the actual farm where many of the ingredients originated is very appealing. These meals are often prepared by a chef with a catering license and almost always include the use of local produce and ingredients. If you are considering farm dinners, start making notes as to what items are available seasonally in your area—this will help you optimize the timing of the events. As mentioned before, regulations will vary, so don't neglect to explore this angle before proceeding. I would recommend starting with the local health department to determine food-handling laws regarding hosting a farm dinner at your location.

A gorgeous location, beautiful presentation, and excellent food make for successful farm dinners at Prairie Fruits Farm in Illinois. Dinners feature the farm's goat cheeses and local produce and meats.

As with any event on your farm where visitors will be present, don't forget to make sure that your liability insurance and coverage will be in effect for any such undertakings.

Farmers' Markets

The number of farmers' markets continues to grow across the country. Larger metropolitan areas usually have several, some of which have multiyear waiting lists for new vendors. Many cheesemakers spend the market season rushing to attend as many as four to six markets *per week*. Family members and other cheesemakers often pitch in to make it possible to have a presence at this many markets.

While many cheesemakers thrive on the interaction with the public that a farmers' market affords, others tire quickly of the routine—and of the mounting work that is left undone on the farm. Here are some suggestions for making sure that your farmers' market experiences benefit your farm's bottom line.

Licenses, Permits, Taxes, and Regulations

First, find out what licenses and/or permits are required to legally sell cheese at markets in your area. The laws vary greatly, from "temporary food market" permits or "traveling retail licenses" to income and sales taxes collected by cities and townships where you are selling, as well as special rules for the handling of "potentially hazardous foods" (PHFs), of which cheese is one. Depending on your location, you might encounter any number of complications and roadblocks for direct sales. The manager of the farmers' market should be well versed in these regulations, but you should be sure to obtain any information from your state and local jurisdictions regarding the regulation of farmers' market sales. In the end, you—not the market manager—will be held responsible.

Budgeting

Once you have researched your jurisdiction's regulations on farmers' market sales, create a projected budget for a full season of selling at each particular market. Be sure to include fuel costs, booth fees, food you might purchase at the market during the day for yourself, and any other costs you can think of. Then project an income from market sales. Talk to other cheesemakers and vendors and get a range of "good day versus bad day" sales to help define your expectations. It's also important to factor in any costs incurred by you being away from the farm, such as hiring extra outside labor to perform tasks that you would otherwise be doing if you were present, or other quantifiable losses. At the end of the season, update the projected budget to reflect your actual income and expenses related to the market.

Laini Fondiller, Lazy Lady Farm, Vermont, selling her cheeses at the Montpelier Farmers' Market. Laini uses a clever display box that holds ice in a lower section and allows the customer to see the cheeses while they are kept cold.

Increasing Income, Reducing Costs

Finally, analyze what you can do to increase your income and decrease the costs:

1. Would cutting out less profitable markets save you money?
2. Are there things you can do to increase sales at market?
   a. Are your display and booth clean and appealing?
   b. Do you provide literature and information that help sell your farm and its products?
   c. Are your products as good as their samples? Don't get lax about quality: Taste every batch, and don't sell any product that is not representative of your usual quality.
   d. Can the customer get your products home in the best condition possible? Consider providing ice packs during the summer (the customer can bring them back the following week) or extra cushioning for especially delicate cheeses.
   e. Do you show up regularly? It doesn't take many missed markets to lose customer loyalty.
3. Would sharing booth space and labor with another producer (if it's allowed) increase profitability?
4. Would adding other farm products (more on that later in this chapter) bring more customers to your booth?
5. Are you selling "yourself"? Don't forget that most customers want to live vicariously through you. Indulge that need by sharing stories, photos of life on the farm, and a newsletter with recent updates (animal births, etc.), and, most important, don't be afraid to open up!

While most cheesemakers assume that they must rely upon farmers' market sales, it is possible for wholesale accounts to be equally profitable once you look at the costs of each.

Increasing/Prolonging Your Cheese Production

Many cheesemakers turn to either buying milk from an outside source (especially during slow production months), milking year-round, or freezing milk or curd during peak production months as a way of increasing the productivity of the facility. Here are some things to think about when considering these options.

Milking Year-Round

There are two methods commonly used to have a year-round milk supply. The most common is to plan birthings for more than one season per year. Another method is to keep animals in milk, without a "dry" period, for more than one year at a time.

Year-round birthing for dairy cattle is a simple matter, but some goats, especially those of European descent, are seasonal breeders, only coming into estrus and able to breed in the fall (for spring birthings). The breeds of African origins, however, are usually capable of naturally cycling into estrus at other times of the year. It is becoming more common to use lighting techniques to encourage off-season fertility in goat herds.

Extended lactations (also referred to as "milking through") means not rebreeding animals annually, but instead continuing to milk them for more than one full milking season. Dairy goats are known to milk for years at a time using this management option. Evin Evans of Split Creek Farm in South Carolina usually breeds her does only two or three times during their entire life span of 10 to 14 years. While does typically drop production during the winter, she says they all come back to close to peak during the spring. The lower production in the winter months is more than made up for by having no dry period or milk lost to feeding kids. This also allows Evin and her team to have far fewer babies to care for and find a place for in new homes. Here at Pholia Farm we always milk several does for at least two continuous years. We choose animals in their second or better lactation. We do not see elevated somatic cell counts for these animals during the winter months, as you would with animals who have been rebred and are toward the end of their lactations. An added bonus for the cheesemaker is a more concentrated milk that is higher in butterfat and protein than at other times of the year.

Dairy cows are rarely milked through, but they are completely capable of doing so. With only one (usually) calf per cow, however, it is often more practical to rebreed; you are dealing with far fewer offspring for the volume of milk, and the need for replacement heifers can drive the decision. (A cow could conceivably have very few opportunities to "replace herself" over her career, especially compared to the number of chances for a dairy goat to produce female offspring.)

MICRODAIRY DESIGNS, LLC

One of the most intriguing setups for mechanically filling and capping fluid milk, yogurt, and other bottled products comes from Frank Kipe's MicroDairy Designs company. In addition to his reasonably priced pasteurizer/vat, he manufactures the EcoFlex Packager, which allows the producer to fill approximately five to ten containers per minute. It can accommodate quite a few varieties of containers and lids. His design goals focus specifically on the very small producer whose volume will be low and who cannot afford the high investment cost of a typical mechanical bottler. The EcoFlex Packager starts at only $1,500. If you don't already have the peristaltic pump that comes with the MicroDairy Designs vat pasteurizer (for more on cheese vats and pasteurizers, see chapter 9), then you will need to purchase that for an additional $1,500.

Frank is well known for his solid customer service after the sale. Lisa Seger of Blue Heron Farm, a small goat cheese dairy in Texas, told me, "Frank is our hero; without him and his products we could not have afforded to farm."

Dairy sheep have shorter lactations than most dairy goats and cows and are never, to my knowledge, milked through.

One caveat to milking through is the anecdotal evidence that fertility is somewhat negatively affected. Some breeders report that animals who have been milked for several lactations can be difficult to settle in subsequent breeding attempts. This is not something we have had an issue with here at Pholia Farm, however.

Buying Supplemental Milk

Many small creameries turn to outside sources for additional cheesemaking milk. Some do this throughout the year, others only during the slower months. For example, a goat or sheep dairy whose own animals might be dry during the winter might purchase cow's milk in order to maximize the use of the creamery equipment year-round. If you can find a local, high-quality source of milk, this can be a viable option for increasing profits. (Cheeses made with this milk will not, of course, be considered farmstead products.)

Appendix D contains a sample milk purchase agreement. You will want to customize such an agreement to make sure that all of the quality issues that are of concern to you are adequately addressed.

In addition to quality, transport of the milk is often an issue. The volume will be too small to warrant any sort of milk tanker truck. Stainless steel milk cans are usually used, although some states allow for the transport of raw milk in food-grade plastic barrels or bags. A milk hauler license is usually required.

EcoFlex Packager by MicroDairy Designs, LLC.

Freezing Milk, Cheese Curd, and Cheese

Another option for prolonging cheese production is the freezing of milk, curd, and finished product (usually soft "fresh" cheese).

Sheep's milk freezes more readily than cow's or goat's milk, with little change to the component structure. In addition, sheep's shorter lactation increases the value of freezing as an option.

Cheese curds, after much of the whey has been released in the vat, can be drained, frozen, and then turned into finished cheese at a later date. Some non-farmstead producers routinely purchase frozen curd from other sources to augment their own production. For the small, artisan cheesemaker, the steps involved require a likely increase in energy usage and storage space, but it can be another way to spread the production out over the year.

Fresh chèvre, fromage blanc, and other soft cheeses are routinely frozen before shipping or simply to keep a stock of product on hand to sell during the period when the animals are dry. As with the other options, additional energy will be utilized, but sometimes being able to maintain a market for your product throughout the year might have a worthwhile payback.

Selling Fluid Milk

While cheese used to be the product that a dairy would turn to in order to add value, for the farm that produces primarily cheese, fluid milk can become a value-added option. There are basically two options when selling fluid milk: first, selling to another cheese or fluid milk producer; and second, selling bottled milk to the end consumer.

For the small cheesemaker, the likelihood of selling excess fluid milk to a larger cheese plant is fairly low. Larger-scale creameries and bottlers usually need a steady, predictable supply of milk. There are smaller creameries, however, that have a regularly scheduled pickup of milk. This can allow for other activities and work to occur on a farm without being, as we say, "a slave to the vat." Milk can also usually be diverted to the fluid producer when you are unable to make cheese for various reasons, such as health or vacation.

For those considering bottling milk themselves, remember that while finished cheese has a much higher price per pound than milk, fluid milk has a lower cost of production and a quick turnaround. Many factors will influence the profit margin, including:

- Cost of equipment for bottling: Some states require mechanical filling and capping. (See the "MicroDairy Designs, LLC" sidebar about small-scale bottling equipment.)
- Pasteurized or raw milk: State laws vary on legal sales of raw milk. In states that allow raw-milk sales, prices will be higher for raw, while your processing cost will be higher for pasteurized milk.

- Price: In some regions people readily pay $8.00 or more per quart for goat's milk and $12.00 or more per gallon for cream-top cow's milk.
- Shelf life: If demand is not high enough, you will lose profit due to the short shelf life of fluid milk in comparison to cheese.

Meat

Most farmers are pretty practical folks. When you raise livestock—even dairy animals—you know that the end of the line for some of them will be the dinner plate. The sale of packaged meat is highly regulated by the USDA as well as state agencies. Slaughter and butchering (cutting and wrapping) of animals whose meat will be sold in grocery stores, distributed, or served in restaurants must occur at USDA or so-called state-equivalent inspection facilities. If you wish to sell individual cuts of meat to customers at your farm or at the farmers' market, you will most likely be required to transport the live animals to one of these inspected facilities for slaughter and butchering. In some states poultry and rabbits are allowed to be processed in on-farm, state-inspected facilities; some states even allow for a limited number of animals to be processed by the farmer, in non-inspected facilities, for direct and farmers' market sales.

The regulation of beef, pork, lamb, and chevon (goat meat) sales when sold direct from the farmer to the end user vary from state to state. Usually animals can be sold in whole or part to private customers, "custom" slaughtered on the farm (which is more humane, as the animals are not stressed from transport or change of environment), and processed by an inspected, custom processor or alternately by the customers themselves. It is assumed that the animal has been inspected and chosen by the customer before slaughter, thereby negating the need of government oversight. Meat processed in this way cannot be resold or used for any commercial purpose and must even be labeled "not for sale."

Lamb and goat meat sales often revolve around ethnic holidays. The Illini SheepNet and Meat GoatNet at www.livestocktrail.uiuc.edu/sheepnet/ has a calendar of these dates as well as other tips for marketing lamb and chevon.

If you should decide to pursue the meat products avenue for your farm, be sure to investigate the legalities and regulations for your area and any venue in which you might want to market your product.

Other Products

Small dairies often produce their own line of milk lotions and soaps as well as confections, fudge and caramels being the two most popular. If you have a farm store or farm stand, or you participate in farmers' markets, then you should be able to realize enough profit from these ventures to make them worthwhile. Additional farm products can also increase customers' interest in your entire product line. (That goes back to the power of brand identity that we talked about in chapter 3.) If you grow fruit trees, produce, herbs, or flowers, these can be sold from the farm, included as ingredients (be sure to understand those good manufacturing processes, or GMPs, we talked about in chapter 12!), or even turned into preserves (with your inspector's approval and/or the right license) and accompaniments for your cheeses.

Evaluating Your Process

There is nothing like routinely putting your process under the microscope and seeking out flaws that are draining your profits. I find that many small cheesemakers don't seem to be willing to take this step, perhaps because most of us are far less drawn to the business side of this life than to the tactile cheese and interactive animal sides. I have always believed that if you are in general happy with your work and able to pay your bills, then that's all well and good—but if you find that no matter how hard you work you are not making ends meet, then it is time to take a good, hard look at your process and take steps to tighten it up.

Most of us cheesemakers love to create new cheeses, which is great—unless you are struggling to survive as a business. Then your passion might need to take a backseat to realism . . . for a while, anyway!

Look for some of these common hidden costs:

1. *Energy*: Look for processes that waste energy, such as long cold storage of product, equipment that is inefficient, batch sizes that are too small, etc. While you may not have an option for some of these costs, they should be kept in mind and readdressed as needed.
2. *Product line:* Look for products that do not sell as well or tend to have a shorter shelf life, thus creating more waste. Consider tailoring your product line to best address sales and profit margin. Most of us cheesemakers love to create new cheeses, which is great—unless you are struggling to survive as a business. Then your passion might need to take a backseat to realism . . . for a while, anyway!
3. *Labor:* Growth does not always bring equivalent profit. Analyze the increased labor costs of producing more product before turning to growth as a means of increasing profit margins.
4. *Herd size:* This goes hand in hand with *labor* and *energy*. The first thing most of us do when looking at ways to increase our profit

is to consider adding more milking animals and therefore creating more cheeses to sell. It may be that this is the right decision, but be sure to look at the increased energy usage in processing the additional milk and cheese, as well as the increased labor costs. In addition, feed costs can be volatile and can greatly affect the production cost of your product. Before you consider growing your herd size, consider . . .

5. *Herd productivity:* Often the herd owner does not spend much time tracking the productivity of individual animals and their offspring. While it is a time-consuming task, this kind of analysis can help you build a herd of efficient, productive animals that also provide you with better-quality replacement milkers. I believe Dairy Herd Improvement (DHI) milk testing to be one of the most practical tools for tracking productivity and health in the dairy herd. As a cheesemaker, it will also allow you to analyze solids—milk fat and protein—for maximum cheese yield, as well as somatic cell count for quality milk and overall udder health in your herd.

With any value-added product, the effort and investment must be worth your while. It may sound easy to add a line of gourmet fudge to your product list, but ingredients, labor, packaging materials, and time will all take away from some other aspect of your business. When you sit down to think about these options, really try to focus on what your original goal was and how these products, events, and opportunities will facilitate that goal. If you find that adding some of these options are valuable from the standpoint of your own mental stimulation and personal gratification, then I say they are worth the time and cost. But more on that in the next and final chapter!

· 14 ·

Keeping the Romance Alive: Tips for Re-Energizing

No matter how great a cheesemaker and herdkeeper you are, sustaining the level of commitment and enthusiasm that is required is, I believe, one of the most difficult parts of this job. Every career has its pressures, but most jobs also include days off and vacation time—even sick time! The farmstead cheesemaker is unlikely to have a union representative showing up to demand better working conditions, so you will need to become your own best advocate.

Lightening the Workload

Cheesemakers around the world have developed some intriguing ways to give themselves breaks from the rigors of dairy work. Any kind of reprieve from work will likely correlate with a loss of potential income, but if that respite means a more sustainable business, then it is probably worth it. You shouldn't always let income define success. As Albert Einstein once said, *"Not everything that can be counted counts and not everything that counts can be counted."*

RULES FOR A SUSTAINABLE WORKLOAD

1. *Don't choose this profession at the wrong time in your life:* If you have too many other life responsibilities—such as caring for young children or elderly parents, diminishing mental and physical energy due to age or illness, or any other burden that will be a major distraction—enter this field with caution.
2. *Don't be a martyr:* Our culture tends to canonize those who are overachievers, work extreme hours, or otherwise deprive themselves of a "normal life" for the sake of their career. But you have to live *this* life: If your cheesemaking career takes you away from commitments and important relationships, then seek a new balance.
3. *Always remember that, no matter how hard you work today, you will never be done:* Find ways to "ignore" the pressure of unfinished projects long enough to stop and smell the cheese curds—and the roses!

Seasonality

One of the most common ways small dairies give themselves a change of pace is by milking and making cheese seasonally. While many people dry their animals off for at least a few weeks in the winter, some people do the opposite. At Orb Weaver Farm in Vermont, Marjorie Susman and Marion Pollack milk their cows only in the winter. During the spring and summer they pursue their other passion and grow produce for market. Here at Pholia Farm we have always milked year-round, but this year we are trying something new. We are milking a few goats through the winter (they will have extended lactations of at least two years before having kids again), and instead of making cheese, we will be freezing that milk and saving it for kids. Since the milk will not be for cheese, we won't have to be as careful and can even have someone come milk for us while our family does something we have never done before: go away, *together!*

Reduced Milking Frequency

In Pat Coleby's book *Natural Goat Care* she lists several ways that small goat dairies create some extra time for themselves, including in Norway milking only one time per day after four or five months of lactation and in France skipping one milking per week (Sunday evening was mentioned). Each of these choices brought decreased production, but they allowed the farmers either a break or the chance to do other high-priority work. Ideally you would have a relief milker; however, availability and cost can be prohibitive for many farmers.

Brian Futhey of Stone Meadow Farm in Pennsylvania milks his sixteen cows once a day for the entire season. This allows him, with minimal assistance, to do all of the milking, make 200 pounds of Camembert and other cheeses per week, and sell the majority of it at four farmers' markets. While he would like to harvest the milk twice a day, he has not found dependable, affordable milkers, and the reality of the farmers' markets means he would have to be in two places at once. He has found that this choice has made his work sustainable.

Flexible Skills

For the very small dairy, having every member of the team able to perform all of the priority tasks can be very helpful. Not only does it ensure continuity—if one member of the team is sick or injured, the most important work will still occur—but it also allows for the possibility of each member being able to get off the farm for short periods of time. Even if the work is not done to the skill level of its usual technician, at least it will get done!

Optimizing the "Break-ation"

Okay, I don't know if we really need another word with "-ation" hyphenated into it, but here it is! Your workday as a farmstead cheesemaker will be long, but you can structure in ritual breaks that you can look forward to and enjoy. Tea time, happy hour, whatever you want to call them: Don't answer the phone, don't think about what you still need to do—just take a break. If you don't make these a ritual, though, you will find yourself making excuses to skip them. Here at Pholia Farm we have happy hour from 4 to 5 p.m. and a weekly breakfast out that we refer to as our "business meeting." We also have "sleep-in Sunday" where we press the snooze button on the alarm once and "tardy Tuesday" when we get up at the regular time but sit with our tea and coffee for 5 or 10 minutes longer than usual. Silly, maybe, but it is rather healthy and fun to be rebellious (even if it is toward yourself) once in a while!

The Day Off

Several farms I talked to have set aside one day of the week for no farm work, other than the essentials. This day, too, needs to become a ritual. Decide whether you will visit family, see friends, or just go to a movie. By setting aside some time for yourselves, your work will be more productive and satisfying during the rest of the week. Remember, there will always be more work that needs to be done, whether you work every day for the rest of your life or not! Don't forget to do the other things that make your life fulfilling and create memories with and for loved ones.

A properly sized goat herd that both the land can sustain and the owners can manage well, Black Mesa Ranch, Arizona.

Stimulating Your Passion
(Yes, you can let the kids read this section!)

While the previous ideas are meant to give you a break from the farming, this section is meant to help you find ways to reconnect with the original passion you felt for cheese, cheesemaking, and the farm life. While many of us find reasons to not participate in some of these activities (they are usually financially based), the value of taking advantage of some of these opportunities cannot really be quantified.

Industry-Related Vacations

The American Cheese Society Conference, Slow Food's "Cheese" event in Bra, Italy, and international cheese tours are just a few of the opportunities for travel related to your business (and hence possible tax deductions). While none of these options could be called economical, their payback can be significant. Connections and relationships are developed that can aid your business in a number of ways, but the immediate reward is the enthusiasm that you bring back to your farm from interacting with others who live a similar life and with those who simply appreciate what you do. In addition, knowledge, ideas, and inspirations are abundant during such events.

Cheese Guilds

The number of formal and informal cheese guilds in the U.S. continues to grow, along with the opportunities and assistance that they provide to their membership. (See appendix A for a list of guilds, current as of this writing.) Several guilds hold annual cheese festivals that showcase their members' products and bring attention to the industry as a whole. Food writers and cheese "celebrities" are often guests at such events, as well as at classes for both guild members and the public. Guild membership and participation is another way to connect with the big picture in your state or region. This connection can help you feel less isolated, and part of an important movement. In addition, the relationship with other guild members can be an asset in tough times. Cheesemakers often rally to support their peers after a disaster or serious setback, not unlike the "old days" when farmers helped each other bring in their crops or raise a barn.

Writing

From blogs to books, many small farmers are finding ways to express themselves and share their lives and passion with others. Both reading and writing about this lifestyle can help you refocus and re-energize. I still remember reading Meg Gregory's (Black Sheep Creamery) blog at her website during the fall of 2007 when disaster struck Meg and Brad's small creamery in southern Washington State. A flood destroyed the majority of their flock, flooded their home and creamery, and floated their cheese-aging trailer around their yard. The blog entries shared their struggles and triumphs in surviving the kind of pain—seeing their beloved flock

Vern Caldwell at cheese sampling event, Foster and Dobbs, Portland, Oregon.

wiped out or sickened—that most of us cannot even imagine. The Gregorys' sharing of their journey was an inspiration to me and put some perspective on the insignificant day-to-day annoyances some of us complain about.

Whatever you want to call it—"belief window" or just "it's all how you look at it"—keeping a perspective that allows you to take minor setbacks and struggles in stride is imperative to surviving as a farmstead cheesemaker. You will need to train yourself to step back mentally and look at the bigger picture of your life. Remind yourself of your original priorities and goals. Take time to enjoy your animals and their antics. Look at your beautiful cheese in its packages, and breathe the sweet, acid smell of fresh curds or newly pressed "infant" wheels. Walk in your fields or forests and see the health of your land. Climb in the hay barn with your kids and see the magic of the life you are giving them.

Then get back to work.

YOU DID IT! You've stuck with it and finished reading the final chapter (of this book, anyway)! I have done my best to offer you all the information and advice I can think of to prepare you to create and sustain the best possible business and long-term future as a farmstead cheesemaker. After everything you've read here, I hope you feel excited and like you can't wait to get going, or a little overwhelmed, or maybe a little bit of both! Just remember:

> *"Obstacles are those frightful things you see when you take your eyes off your goal."*
> —Henry Ford, 1863–1947

APPENDIX A

Resources

This list is not meant as an endorsement by the author or the publisher of any of the companies, organizations, or individuals mentioned. Companies may carry more supplies and equipment than those listed. They also may have used and/or reconditioned equipment for sale.

Books, Publications, Resources for Learning*

American Farmstead Cheese: The Complete Guide to Making and Selling Artisan Cheeses, by Paul Kindstedt (Chelsea Green Publishing, 2005). *Intermediate to advanced cheesemaking and business. This book has successfully bridged the gap between industrial cheesemaking tomes and hobbyist/home cheesemaking books. Highly recommended.*

Cheese Problems Solved, edited by P. L. H. McSweeney (CRC Press, 2007). *A serious textbook (and priced like a textbook) that is very technical and totally stimulating for cheesemakers who just have to know more than maybe they should.*

The Cheesemaker's Manual, by Margaret P. Morris (Glengarry Cheesemaking and Dairy Supply, 2003). *Great book for those transitioning from hobbyist to commercial.*

Code of Federal Regulations (CFR), Title 21, Part 110. www.fda.gov. In the A–Z Subject Index bar across the top of the webpage, click on C and then click on Code of Federal Regulations. In the box labeled Search CFR Title 21 Database, type in 110. (Or search "Code of Federal Regulations (CFR), Title 21, Part 110" on the FDA website.) *While this is not an easy-to-read document, you should at least know how to access the CFR if needed. The Dairy Practices Council handbooks (below) reference the CFR when referring to GMPs.*

CreamLine. www.smalldairy.com. *Quarterly publication by longtime small-dairy advocate and cheesemaker Vicki Dunaway.*

Dairy Practices Council, Educational Guidelines for the Dairy Industry (Newtown, PA). 215-860-1836. www.dairypc.org. *One of the best resources for facts on all aspects of the dairy industry: HACCP, construction, quality milk production, equipment, etc. Purchase booklets individually or in binders related to your production.*

The Fabrication of Farmstead Goat Cheese, by Jean-Claude Le Jaouen (Cheesemaker's Journal, 1990). *Superb book for those whose focus will be soft-ripened goat cheese.*

Goat Cheese: Small-Scale Production, by the Mont-Laurier Benedictine Nuns (New England Cheesemaking Supply Co., 1983). *Another great little book focused on soft-ripened goat cheese production.*

*There are other books and publications available that either I have not read or are more suitable for beginning or industrial cheesemaking. This list comprises those I have read and use myself.

Cheesemaking Supplies: Forms, Cultures

Dairy Connection (Madison, WI). 608-242-9030. www.dairyconnection.com. *Culture selection specialists.*
Fromagex (Rimouski, Quebec, Canada). 866-437-6624. www.fromagex.com (click on English for language selection). *Forms and equipment.*
Glengarry Cheesemaking and Dairy Supply (Lancaster, Ontario, Canada). 888-816-0903. www.glengarrycheesemaking.on.ca. *Equipment, supplies, information, and advice for cheesemakers.*
New England Cheesemaking Supply (Ashfield, MA). 413-628-3808. www.cheesemaking.com. *Supplies for the home cheesemaker.*

Classes and Instruction

University

California Polytechnic State University ("Cal Poly"), Dairy Science (San Luis Obispo, CA). 805-756-1111. www.calpoly.edu. *Short courses in cheesemaking.*
Oregon State University, Food Science and Technology. 541-737-8322. http://oregonstate.edu/dept/foodsci/extservices/ext_index.htm. *Respected program in OSU's remodeled and expanding dairy facility.*
University of Guelph, Dairy Science and Technology (Guelph, Ontario, Canada). 519-824-4120. www.foodsci.uoguelph.ca/dairyedu/home.html. *One of the oldest cheesemaking short courses in North America.*
University of Vermont, Vermont Institute for Artisan Cheese (Burlington, VT). 802-656-8300. www.nutrition.uvm.edu/viac. *Several highly respected short courses throughout the year.*
University of Wisconsin–River Falls, Dairy Science (River Falls, WI). 715-425-3911. www.uwrf.edu. *Short courses that apply toward Wisconsin's cheesemaker requirements.*
Utah State University, Western Dairy Center (Logan, UT). 435-797-3466. www.usu.edu/westcent. *A large variety of cheesemaking and related courses.*
Washington State University, WSU Creamery (Pullman, WA). 800-457-5442. www.wsu.edu/creamery. *Respected short courses since 1986. Makers of Cougar Gold cheese.*

Private

Glengarry Cheesemaking and Dairy Supply (Lancaster, Ontario, Canada). 888-816-0903. www.glengarrycheesemaking.on.ca. *Workshops in Ontario and guest teaching by Margaret Morris.*

Neville McNaughton (Davisville, MO). 314-409-2252. www.cheezsorce.com. *Great reputation for instruction. Also producing some small cheese vats. Website a bit lacking.*

New England Cheesemaking Supply (Ashfield, MA). 413-628-3808. www.cheesemaking.com. *On-staff technical support. Website has lots of good photos of cheeses being made and more.*

Peter Dixon (Westminster West, VT). 802-387-4041. www.dairyfoodsconsulting.com. *Great reputation for teaching and consulting. Website has up-to-date class lists, recipes, and more.*

Cleaning Chemicals, Supplies

All QA Products (Mount Holly, NC). 704-829-6600. www.allqa.com. *Sanitizer strength test kits.*

Ecolab (worldwide). U.S. customer service: 651-293-1963. www.ecolab.com. *Large supplier of food industry cleaning products. Website is a bit awkward to navigate.*

PureLabs (Madison, WI). 608-316-3500. www.purelabs.com. *Detergent chemistry.*

Energy: Grants, Renewables

Dairy Farm Energy (Ithaca, NY). 607-266-6401. www.dairyfarmenergy.com. *Controlling energy costs.*

Database for State Incentives for Renewables and Efficiency (DSIRE). www.dsireusa.org. *Information on state, local, utility, and federal incentives and policies that promote renewable energy and energy efficiency.*

Equipment: Milking Parlor, Milkhouse

Coburn Company (Whitewater, WI). 800-776-7042. www.coburn.com. *Wholesale distributor of dairy supplies and equipment.*

Glengarry Cheesemaking and Dairy Supply. (Lancaster, Ontario, Canada). 888-816-0903. www.glengarrycheesemaking.on.ca. *Small imported bulk tanks.*

Hamby Dairy Supply (Maysville, MO). 816-449-1314. www.hambydairysupply.com. *My favorite place for milking equipment, milk cans, supplies. Can help with problems with any type milking system—even those they don't sell.*

Major Farm/Vermont Shepherd (Putney, VT). 802-387-4473. www.vermont shepherd.com/headgates/about.html. *Cascading headgates, platforms, automatic feeders, and ramps for goats and sheep.*

Nasco Farm and Ranch (Fort Atkinson, WI, and Modesto, CA). 800-558-9595. www.enasco.com. *Dairy and livestock supplies. Great catalogs. Sells Coburn Company milking equipment.*

Parts Department Dairy Equipment and Supplies (Sharon, CT). U.S.: 800-245-8222; outside U.S.: 860-364-9326. www.partsdeptonline.com. *Milking equipment and replacement parts. Awkward website, but good customer service.*

Recorder Charts and Pens (Newhall, CA). 800-758-0740. www.recordercharts andpens.com. *Recording chart paper, recorder pens, marking system supplies.*

Equipment: Make Room

C. van 't Riet Dairy Technology USA (DuBois, PA). 814-591-6979. www.schuller.us. *Vats, pasteurizers, presses, forms, more.*

Central Restaurant Products (Indianapolis, IN). 800-215-9293. www.central restaurant.com. *Shelving (including MetroMax), carts, prep tables, and more.*

Dairy Heritage (Hagerstown, MD). 301-223-6877. www.dairyheritage.com. *Vats, bottling equipment, more.*

DR Tech (Grantsburg, WI). 715-463-5216. www.drtechinc.net. *Manufacture, sales, and service of large-scale mechanical processing systems for the dairy industry. Also makes smaller custom vats and products.*

Glengarry Cheesemaking and Dairy Supply (Lancaster, Ontario, Canada). 888-816-0903. www.glengarrycheesemaking.on.ca. *Presses, vats, pasteurizers.*

JayBee Precision (Dayton, OH). 800-236-3956. www.thevatpasteurizer.com. *15- and 30-gallon vat pasteurizers.*

Kleen-Flo Dairy Equipment (Lynn, IN). 765-874-1292. www.kleenflo.us. *Small pasteurizer/vat combos, round and square vats, butter churns. Website does not show all equipment; check their price list for choices.*

Kusel Equipment Co. (Watertown, WI). 920-261-4112. www.kuselequipment.com. *Larger vats, but also "lab size/small production" vats starting at about 65 gallons.*

MicroDairy Designs (Smithsburg, MD). 301-824-3689. www.microdairy designs.com. *Small, more affordable pasteurizer. EcoFlex small bottling system for milk, yogurt, etc.*

Qualtech (Quebec, Canada). 888-339-3801. www.qualtech.ca (click on English). *Makes small and large vats (mixed reviews from cheesemakers as to ease of initial setup and service).*

Stanpac (Lewiston, NY; Dallas, TX). 905-957-3326. www.stanpacnet.com. *Glass milk bottles.*

Equipment: Aging Room

Beverage Factory (San Diego, CA). 800-710-9939. www.beveragefactory.com. *Wine cellar cooling units by many manufacturers.*

Burch Industries (Laurinburg, NC). 910-844-3688. www.burchindustries.com. *Egg coolers.*

Central Restaurant Products (Indianapolis, IN). 800-215-9293. www.centralrestaurant.com. *Shelving (including MetroMax), carts, prep tables, and more.*

CoolBot (New Paltz, NY). 888-871-5723. www.storeitcold.com. *Temperature control units for converting window air conditioner to aging room cooler.*

Lab Equipment, Supplies, and Miscellaneous

All QA Products (Mount Holly, NC). 704-829-6600. www.allqa.com. *Sanitizer strength test kits, thermometers, pH strips.*

Cole-Parmer (worldwide). 800-323-4340. www.coleparmer.com. *pH meters.*

Nelson-Jameson (Marshfield, WI). 800-826-8302. www.nelsonjameson.com. *Delvotest P test kits, lab supplies, cleaning supplies, and more—request a buyers' guide.*

Professional Equipment (Janesville, WI). 800-334-9291 www.professionalequipment.com. *Digital thermometers, hygrometers, Kill-a-Watt meter (for energy usage tracking).*

Laboratories for Milk and Product Testing

For a complete list of labs approved for testing milk for interstate shipping, go to **www.fda.gov** and follow the links (this is a pretty unwieldy website, but stick with it!) for Food/Food Safety/Product-Specific Information/Milk Safety/Federal State Programs/Interstate Milk Shippers List. *Most of these labs are private, but those that do outside testing are listed as well.*

Agri-Mark Central Laboratory (West Springfield, MA). 413-733-6213. www.agrimark.net/public/facilities.php. *Lab offers outside testing for milk and product, pathogens, and components. Parent company of Cabot Creamery and others.*

Silliker (worldwide). www.silliker.com. *Pathogen testing for milk and product.*

University of Minnesota Veterinary Diagnostic Laboratory, Laboratory for Udder Health (St. Paul, MN). 612-625-8787. www.vdl.umn.edu/ourservices/udderhealth/home.html. *Tests related to udder health.*

Major Lending Institutions/Organizations

Commercial Lending Institutions
Banks, credit unions. Direct loans. Working with your local bank can be an advantage if you already have a good relationship with it and if the bank has a local focus and community-building reputation.

Farm Services Agency (FSA)
www.fsa.usda.gov

Part of the USDA. Direct and guaranteed. Beginning farmer and socially disadvantaged (formerly called minorities) programs. Low interest rates. Visit the FSA website or contact your local office. Other financial assistance programs are available through conservation, energy programs, and disaster assistance.

Farm Credit System (FCS)
www.farmcreditnetwork.com

A national network of cooperative lending institutions in existence since 1916. Borrowers become owners of the FCS. Long- and short-term direct loans as well as revolving credit. Visit the FCS website to locate a local office.

Certified Development Company (CDC)
www.nadco.org

Regional organizations that promote and assist business development through loans and other programs. Interest rates usually higher than FSA loans. For a state-by-state list of CDCs, go to the National Association of Development Companies website.

Small Business Administration (SBA)
www.sba.gov

Direct and guaranteed through commercial lending institutions, CDCs, and non-profit community micro-lenders. Visit the SBA website or contact your local office. Interest rates vary based on loan program, loan size, and current prime interest rate, but are usually higher than with FSA loans. Micro-lender list on the SBA website; search by state.

NLPA Sheep and Goat Fund
www.nlpa.org

The National Livestock Producers Association (NLPA) was formed in 1921 to assist livestock producers with marketing issues. In the 1930s it added capitalization through livestock credit corporations to its purpose. The sheep and goat fund is specifically targeted at funding projects that focus on developing products from, of course, sheep and/or goats. Loans are not made to individuals (sole proprietors; see chapter 4 for more on sole proprietorship and other business structure options).

Packaging Supplies and Equipment

Cold Ice (Oakland, CA). 800-525-4435. www.coldice.com. *Ice blankets, ice gel packs, temperature indicators.*
Doug Care Equipment (Springville, CA). 559-539-3076. www.dougcare.com. *Vacuum-sealing equipment and supplies.*
Uline (Waukegan, IL). 800-958-5463. www.uline.com. *Boxes, packing material, ice gel packs.*

Cheese Guilds and Associations

American Cheese Society. www.cheesesociety.org
California Artisan Cheese Guild. www.cacheeseguild.org
Maine Cheese Guild. www.mainecheeseguild.org
New York State Farmstead and Artisan Cheese Makers Guild. www.nyfarmcheese.org
Ontario Cheese Society. www.ontariocheese.org
Oregon Cheese Guild. www.oregoncheeseguild.org
Pennsylvania Farmstead & Artisan Cheese Alliance. www.pacheese.org
Raw Milk Cheesemakers' Association. www.rawmilkcheese.org
Southern Cheesemakers' Guild. www.southerncheese.com
Vermont Cheese Council. www.vtcheese.com
Wisconsin Specialty Cheese Institute. www.wisspecialcheese.org

APPENDIX B
Floor Plans

Mama Terra Micro Creamery

Designer: Robin and Gabe Clouser.
Dimensions: 12' × 30'.
Square Footage: 360.
Purpose: Production of fresh, pasteurized cheese.
Capacity: 12–25 does.
Notes: Well-oriented plumbing and drainage. Simple, efficient design. Would benefit from addition of changing/entry space at exterior wall by make room door. (Owners currently enter through milkhouse, causing them additional sanitation concerns.) Footbaths placed outside rooms would help limit concerns.

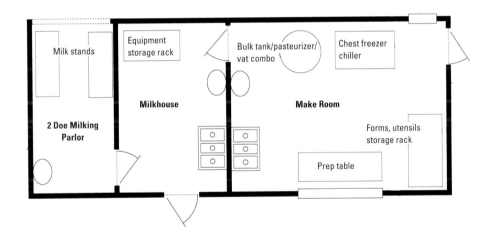

Small Creamery

Designer: Gianaclis Caldwell.
Dimensions: 24' × 36'.
Square Footage: 846.
Purpose: Creamery for production of fresh and aged cheeses. On-farm retail and shipping.
Capacity: 20–50 does or ewes, 5–10 cows.
Notes: Designed for efficiency and functionality. Passive solar orientation and fenestration. On-demand hot water heater.

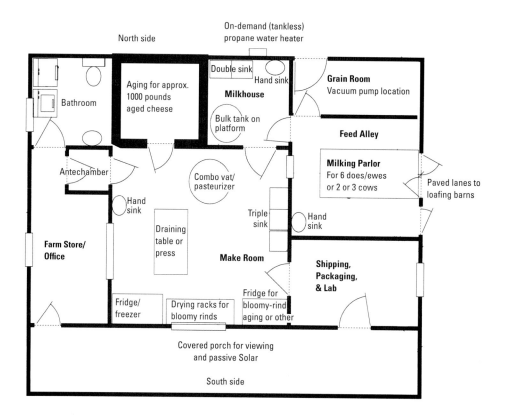

Pholia Farm Barn and Creamery

Designer: Gianaclis Caldwell.
Dimensions: 60' × 60' (*note:* room dimensions are approximate).
Square Footage: Dairy/creamery, about 938; additional unconditioned utility, about 288.
Purpose: Production of aged, raw-milk cheeses. On-farm retail and shipping.
Capacity: 30–50 does (Nigerian Dwarf goats), aging space for 1,400 pounds of cheese.
Notes: I love our dairy, but if I had to do it over, I would make the milking parlor larger and include a packaging and shipping room off the creamery—perhaps utilizing part of what is now the back covered porch. We use the loft of the barn for partial living space and dry storage of boxes, packing materials, etc.

Small Non-Licensed Home Dairy

Designer: Gianaclis Caldwell.
Dimensions: 16' × 16'.
Square Footage: 256.
Purpose: Production of dairy products for home use and private milk sales where legal.
Capacity: 10–20 does or ewes, 1–3 cows.
Notes: Plumbing on one interior wall for cost savings and freeze protection. Floor drains could be included for improved functionality and possible future licensing. Size can easily be altered (maintain 4'-interval dimensions for less building material waste and cost). Small on-demand hot water heater.

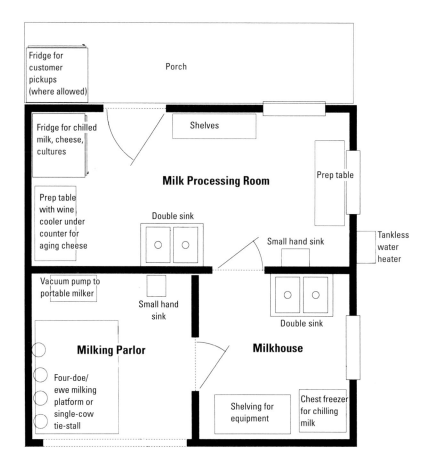

APPENDIX C

Milk and Cheese Quality Tests and Parameters

Appendix C provides an overview of some of the more common tests and how to use the data to your advantage. In addition, we will cover some other avenues for monitoring the production of high-quality milk and cheese. For a more thorough understanding of these tests, please refer to the notes/references section.

Common Laboratory Milk Tests Defined

Standard Plate Count (SPC): Raw milk, total number of aerobic bacteria. Expressed in number of colony-forming units per milliliter of milk (cfu/ml).

Why it matters for cheese: High bacteria counts in raw milk indicate possible mastitis, poor udder preparation, dirty equipment, and/or inadequate cooling. Coliform bacteria (of which *E. coli* is one type) are often the dominant microorganism of the SPC. See below.

Preliminary Incubation Count (PIC): Milk sample is held at 55°F (12.8°C) for 18 hours, then an SPC is performed. Bacteria associated with high PICs are generally *psychrotrophic* (cold-loving).

Why it matters for cheese: Many psychrotrophic bacteria produce *proteolytic* (breaks down proteins) and *lipolytic* (breaks down fats) enzymes that can change the structure of milk during cold storage prior to cheesemaking, leading to poor coagulation, undesirable flavors, and rancidity in the cheese.

Laboratory Pasteurization Count (LPC; also known as thermoduric bacteria count): Milk sample is heated to 145°F (62.9°C) for 30 minutes, then an SPC is performed. Counts bacteria that survive pasteurization. No legal limit for raw milk.

Why it matters for cheese: High LPCs are associated with poor udder cleaning and/or dirty equipment, practices that reduce the quality of milk.

Coliform Bacteria Count: Milk sample is plated on growth medium and incubated at 90°F (32°C) for 24 hours. If total coliform count is high, a separate *E. coli* culture is done to detect pathogenic strains.

Why it matters for cheese: Coliforms from fecal contamination can be pathogenic, causing severe illness and death in certain cases. Coliforms also cause early gas blowing in cheese.

TABLE AC-1: Common Laboratory Test Levels

Test	Milk Test Result Ranges			
	Ideal	Acceptable	Common Industry*	Regulatory Limit
SPC	<1,000/ml	<5,000/ml	<10,000/ml	100,000/producer; 300,000/commingled tanker
PIC	Should be <3–4x SPC	<25,000/ml	25,000–50,000/ml	No legal limit
LPC	<250–300/ml			20,000/ml (pasteurized milk)
Coliform	<10/ml	<50–100/ml		<10/ml (pasteurized milk)
SCC cows/sheep**			>200,000/ml	>750,000/ml
SCC goats				>1,000,000/ml

Source: Dairy Practices Council.
*Related to premium payments within industry.
**Sheep values are currently the same as for cows but are being reviewed.

TABLE AC-2: Possible High Count Causes

Test Result	Natural Bacteria	Mastitis Bacteria	Dirty Udders	Dirty Equipment	Poor Cooling
SPC > 10,000	Doubtful	Possible	Possible	Possible	Possible
SPC > 100,000	Doubtful	Possible	Doubtful	Likely	Likely
LPC > 300	Doubtful	Doubtful	Possible	Likely	Doubtful
PI high, 3–4x SPC	Doubtful	Doubtful	Possible	Likely	Likely
SPC high > PI	Doubtful	Presumed	Doubtful	Doubtful	Doubtful
Coli Count > 100	Doubtful	Possible	Possible	Possible	Doubtful

Source: Bacteria Counts in Raw Milk, Richard L. Wallace, University of Illinois Extension.

Somatic Cell Count (SCC): Counts *leukocytes* (white blood cells) and other body-tissue cells in milk. Indicative of udder health—injury or infection, and/or stage of lactation.

Why it matters for cheese: The increase of SCC correlates with the increase of proteolytic enzymes in the milk, leading to lower cheese yield, poor coagulation, and off-flavors in cheese.

Milk Urea Nitrogen (MUN): Excess protein in diet leads to high blood and milk urea nitrogen levels, which affects cheese quality, as well as contributing to excess nitrogen in manure and ammonia in urine. Acceptable levels for cows average 10–14 mg/dl. Goat and sheep norms have not been established. Small sampling has suggested normal levels for goats to be higher than that of cows.

Why it matters for cheese: High MUN levels are associated with poor coagulation, early fermentation, and excessive proteolysis in aging cheese.

Testing Frequency

The frequency at which your milk will be tested by regulatory officials will vary. I recommend augmenting the state's testing to keep the frequency to every six to eight weeks, minimum. More frequent testing should occur if the results are less than favorable. The important thing is to *establish baseline numbers* that you understand, accept, and can maintain. By watching results closely, you will be able to see any changes that indicate a flaw in your sanitation and practices and then remedy that flaw before it becomes a hazard and/or violation.

Most states perform only the minimum tests required to ensure safe milk. For cheesemaking *quality* (beyond safety), you may want to have other tests performed on a routine basis or when needed. See appendix A for a list of laboratories.

On-Farm and Creamery Tests and Observations

Tests for SCC (Somatic Cell Count) and Mastitis in Milk

Several immediate tests can be done on milk either during milking or prior to cheesemaking that will indicate the presence of somatic cells (SCs) and/or mastitic milk. The California Mastitis Test (CMT) uses a reagent combined with an equal amount of milk swirled in the provided paddle to detect the presence of SCs. The paddle is swirled for 15 seconds and then tilted back and forth. Thickening is judged in degrees that correlate with the amount of SCs present.

- Trace gelling = 300,000
- Thickening but not clumping = 500,000–1,000,000
- Thickening and clumping = >1,000,000
- Clumping and sticking to paddle = >2,000,000

Mastitic milk will be more alkaline than healthy milk. (Some on-farm SCC tests are actually checking pH.) If the pH of milk at the start of cheesemaking is higher than normal, perform a CMT and compare.

Dairy Herd Improvement Programs

Herds enrolled in dairy herd improvement programs have milk samples pulled and weights recorded monthly on each animal. Data collected includes (this can vary by lab and program) butterfat, protein, solids-not-fat, cheese yield (based on butterfat-to-protein ratio and solids), and somatic cell count. This can greatly help the farmer-cheesemaker track seasonal changes in milk and milk quality as they relate to cheesemaking. Many new programs allow for owners to collect samples and data without needing an outside trained test person.

Appearance, Odor, and Flavor

Many flaws and problems can be detected in milk using your senses. I greatly encourage artisan cheesemakers to train their eyes, noses, and taste buds to be their first line of defense in ensuring day-to-day quality.

Tips for Analyzing Milk
1. For maximum aroma analysis, pour sample into a jar, leave a headspace, close tightly, and warm to about 60°F (in water bath). Open jar directly under your nose and inhale deeply.
2. Compare fresh milk to that which has been stored for several days to help pinpoint cause.
3. If necessary, compare samples from each animal to rule out individual variation in flavor.
4. Official advice encourages the pasteurization of all milk samples prior to tasting.

Spoiled or unclean (dirty, animal taste, spoiled, fruity): Contamination and/or bacterial growth during storage. Causes: dirty equipment or udders, psychotrophic bacteria.

Rancid, bitter (soapy, baby vomit, blue cheese): Breakdown of fat. Causes: excess agitation or foaming of milk during milking, late lactation (high SCC), fluctuation of temperature during cold storage.

Malty, acidic (malt cereal and/or sour): Bacterial growth during storage. Causes: poor cooling of milk.

Oxidized (cardboardy, old oil): Oxidation of butterfat. Causes: Excessive high-fat feeds, low levels of vitamin E, presence of certain metals (copper contamination, rust).

Feedy (aromatic, unnaturally sweet): Odors and flavors absorbed by animal and transmitted to milk.

Common Lab Tests for Cheese
These tests are just the tip of the iceberg in regard to understanding the safe manufacture of cheese. I recommend further study through cheesemaking books, classes, and the Dairy Practices Council's Guideline No. 100: "Food Safety in Farmstead Cheesemaking."

TABLE AC-2: Cheese Pathogen Test Levels			
Test	**Ideal**	**Presidium***	**Regulatory Limit**
Salmonella	Negative	Negative	Positive
Listeria monocytogenes	Negative	Negative	Positive
Total Coliforms			
E. coli O157:H7	<10 cfu/g	<100 cfu/g	>10,000 cfu/g
Staph aureus	<10 cfu/g	<100 cfu/g	>10,000 cfu/g
Staph enterotoxin	Negative	Negative	Positive
*Slow Food USA: American Raw Milk Cheese Presidium, Protocol for Presidium Members			

In-House Observations and Tips

1. Product testing can be infrequent as compared to milk and environmental testing. Good manufacturing practices and quality assurance are more important than testing for ensured quality.
2. Test more "dangerous" cheeses—high-moisture cheeses, such as Brie, Camembert, washed-rind types—more frequently.
3. Suspect any finished product that deviates from flavor, aroma, and appearance profiles.
4. During production, flag any batches where the make deviates from normal production standards. Test prior to selling.

APPENDIX D

Sample Milk Purchase Agreement

This agreement is entered into on _____[Date] between _____[Buyer] and _____[Seller].

1. This agreement shall be effective from _____[Date] through _____[Date]. The parties named above may, however, elect to renew this agreement for another term.
2. Buyer agrees to purchase from seller _____[Species] milk not older than ____ days, to be used for the manufacture of cheese for the length of this agreement.
3. Either a) Buyer agrees to pick up milk at the farm, or b) Seller agrees to deliver milk to Buyer at a cost of $_____ per delivery.
4. Seller warrants that milk sold to Buyer shall be free of inhibitory substances and shall meet the standards set forth by the _____ Department of Food and Agriculture for manufacturing of Grade _____ milk. Seller shall remain under State inspection throughout the life of the contract.
5. Milk must meet the approval of the Buyer based on any or all of the following tests:
    a) Taste and odor
    b) Bacteria*
        i. Coliforms not to exceed _____/ml (_____ to_____/ml payable at 90%; _____ to _____/ml payable at 80%)
        ii. SPC not to exceed _____/ml (_____ to_____/ml payable at 90%; _____ to _____/ml payable at 80%)
        iii. LPC not to exceed _____/ml (_____ to_____/ml payable at 90%; _____ to _____/ml payable at 80%)
        iv. PI not to exceed _____% of SPC (_____ to_____% payable at 90%; _____ to _____% payable at 80%)
        v. Other …
    c) Received at no less than 40°F (34°F preferred)
    d) Sediment (per dairy inspector's test)
    e) Any counts over the maximum high count agreed upon are subject to rejection by the Buyer. In the event of a high count that does not change product quality, the first two deductions taken will be 10%.
6. Weekly costs of testing milk and shipping to lab shall be divided equally between Buyer and Seller. Frequency: Milk will be tested _____ by an outside lab and sent in by Buyer. If milk quality problems exist and more frequent testing is needed, Seller will pay costs and be responsible for shipping

samples and proving to Buyer that milk meets quality standards before Buyer purchases milk again. Failure to produce milk that meets quality standards is grounds for cancellation of contract.

7. Buyer shall make no payment to Seller for milk that does not meet the conditions of paragraphs 4 and 5 above. Any payments made prior to testing the milk shall be credited to Buyer if milk does not meet same conditions.
8. If milk is shipped with inhibitory substances, or if milk is of such poor quality that cheese does not set up, the producer of that milk shall be financially liable for actual costs incurred by Buyer of labor, utilities, ingredients, transportation, and any other milk that was contaminated. Costs will be deducted from future milk checks. Seller will provide Buyer proof of liability insurance.
9. The price of milk will be based on a combination of butterfat, protein, and the time of year produced. The butterfat percentage × 0.66 + the protein percentage × 1.33 will yield a number which correlates to the payment schedule agreed upon. Buyer will make every effort to purchase all milk produced by Seller, but cannot guarantee to purchase more than Seller shipped in Seller's lowest quarter of the previous year. Buyer agrees to purchase all milk produced from Seller before adding new producers or purchasing from other outside sources. Seller agrees to give first option for purchase to Buyer for all marketable milk.
10. Buyer will pay a bonus of 10 percent for milk that has lab counts for paragraph 5b, i–iii, at half or less. This means LPC and coliform counts less than _____ and raw counts less than _____.
11. Buyer agrees to renew this contract next year and purchase milk from Seller for another year provided Seller produces quality milk and similar amounts as in the previous year's lowest quarter.
12. Buyer and Seller agree that this agreement may be suspended in the event of Acts of God or circumstances beyond the control of either party. Specifically, the agreement shall be suspended if Buyer dies or becomes disabled or incapacitated, either mentally or physically, so as to be unable to operate his/her business, or if the plant is partially or totally destroyed so as to cause a halt in the business. In addition, the agreement will be suspended if the laws of any governing body prohibit Buyer from manufacturing cheese.
13. Buyer and Seller agree that this agreement shall be nontransferable by either party without the written consent of both parties.

Attach agreed-upon milk purchase base price and payment schedule.

*Bacteria counts left blank, as some agreements are for Grade A and others for Grade B milk.

Contract adapted from one provided courtesy of Sarah Shevett, California.

Notes/References

There are other books and publications available that either I have not read or are more suitable for beginning or industrial cheesemaking. This list comprises those I have read and use myself.

General
Grade "A" Pasteurized Milk Ordinance. U.S. Department of Health and Human Services. Go to www.fda.gov and follow the links (this is a pretty unwieldy website, but stick with it!) for Food/Food Safety/Product-Specific Information/Milk Safety/ National Conference on Interstate Milk Shipments (NCIMS) Model Documents—then click on the most recent PMO.

How to Become a Dairy Artisan. Wisconsin Dairy Artisan Network. www.wisconsindairyartisan.org.

"The Land of Cheese Marketing: Don't Get Caught without a Paddle," by David Major. *CreamLine,* Issue No. 10, Summer 2001. www.smalldairy.com.

On-Farm and Small-Scale Dairy Products Processing. Dairy Practices Council Guideline No. 90. www.dairypc.org.

Chapter 1
Cheeses of the World: An Illustrated Guide for Gourmets, by Bernard Nantet (Rizzoli, 2005).

Eating in America, by Waverly Root and Richard de Rochemont (Ecco, 1981).

International Dairy Foods Association. www.idfa.org.

The New American Cheese: Profiles of America's Great Cheesemakers and Recipes for Cooking with Cheese, by Laura Werlin (Stewart, Tabori & Chang, 2000).

Understanding Dairy Markets. University of Wisconsin Dairy Marketing and Risk Management Program. http://future.aae.wisc.edu/.

Chapters 3 and 4
"Building a Business Plan for Your Farm: Important First Steps," by Rodney Jones. 2003 Risk and Profit Summer Conference, August 14–15, 2003, Manhattan, KS.

Creating a Business Plan for Your Dairy Business, by Russ Giesy. University of Florida Extension. http://dairy.ifas.ufl.edu.

Form Your Own Limited Liability Company, by Anthony Mancuso (Nolo, 2009).

"Sampling and Farm Stories Prompt Consumers to Buy Specialty Cheeses," by Barbara A. Reed and Christine M. Bruhn. *California Agriculture,* Volume 57, Number 3.

Starting and Running Your Own Small Farm Business, by Sarah B. Aubrey (Storey, 2007).

Chapter 5
The Agricultural Employer's Tax Guide. Internal Revenue Service Publication No. 51. www.irs.gov/pub/irs-pdf/p51.pdf.

Are Your Internships Legal? *Growing for Market,* September 2007. www.growingformarket.com.

Exploring the Legal and Ethical Issues of Internships. *Growing for Market,* October 2007. www.growingformarket.com.

Food Insurance. Professional and Liability. www.professional-and-liability.com/food-insurance.html.

Internships in Sustainable Farming: A Handbook for Farmers, by Doug Jones (Northeast Organic Farming Association of New York, 1999). http://nofany.org/publications.html.

Chapter 7

Best Management Practices to Handle Dairy Wastes, by Ted W. Tyson and B. R. Moss. Alabama Cooperative Extension System Publication No. ANR-970. www.aces.edu/pubs/docs/A/ANR-0970/ANR-0970.pdf.

"Creamery Waste Management," by Vicki Dunaway. *CreamLine,* Issue No. 32, Fall 2007. www.smalldairy.com.

Dairy Processing Methods to Reduce Water Use and Liquid Waste Load, by Kent D. Rausch and G. Morgan Powell. Kansas State University–Manhattan Cooperative Extension Service Publication No. MF-2071, 1997.

"Environmental Costs of Home Construction Lower with Wise Choice, Reuse of Building Materials." Consortium for Research on Renewable Industrial Materials (CORRIM) news release, August 24, 2004.

Milking Center Wastewater. Dairy Practices Council Guideline No. 15. www.dairypc.org.

Potable Water on Dairy Farms. Dairy Practices Council Guideline No. 30. www.dairypc.org.

Chapters 8 and 9

The Design, Installation, and Cleaning of Small Ruminant Milking Systems. Dairy Practices Council Guideline No. 70. www.dairypc.org.

Grade "A" Pasteurized Milk Ordinance, 2003 Revision. U.S. Department of Health and Human Services. Go to www.fda.gov and follow the links (this is a pretty unwieldy website, but stick with it!) for Food/Food Safety/Product-Specific Information/Milk Safety/National Conference on Interstate Milk Shipments (NCIMS) Model Documents—then click on the 2003 PMO.

Mastitis Case Studies: Milking Machine. University of Illinois. http://classes.ansci.illinois.edu/ansc438/Mastitis/milkmachine.html.

Maximizing the Milk Harvest: A Guide for Milking Systems and Procedures, by the Milking Machine Manufacturers Council of the Equipment Manufacturers Institute. www.partsdeptonline.com/maximizing_the_milk_harvest.htm.

Milk for Manufacturing Purposes and Its Production and Processing: Recommended Requirements (U.S. Department of Agriculture, Agricultural Marketing Service, Dairy Programs, September 2005). Go to www.usda.gov and search "Milk for Manufacturing."

Milking Parlor Types, by Douglas J. Reinemann (University of Wisconsin–Madison Milking Research and Instruction Laboratory, December 2003). www.uwex.edu/uwmril.

Production and Regulation of Quality Dairy Goat Milk. Dairy Practices Council Guideline No. 59. www.dairypc.org.

Chapter 10
"Cheese Aging Rooms," by Vicki Dunaway. *CreamLine,* Issue No. 11, Fall 2001. www.smalldairy.com.

Current Options in Cheese Aging Caves: An Evaluation, Comparison, and Feasibility Study, by Jennifer Betancourt with help from Amanda DesRoberts. www.silverymooncheese.com.

"A Straw Bale Aging Room," by Larry Faillace. *CreamLine,* Issue No. 11, Fall 2001. www.smalldairy.com.

Chapter 12
Conducting and Documenting HACCP: Principle Number One: Hazard Analysis. Dairy Practices Council Guideline No. 92. www.dairypc.org.

Conducting and Documenting HACCP: SSOPs and Prerequisites. Dairy Practices Council Guideline No. 91. www.dairypc.org.

Cookie Cutter Procedures, by Dominique Delugeau, DCI Cheese Co. Presented at the 2005 American Cheese Society Conference, July 20–23, 2005, Louisville, KY.

Developing a HACCP Program, by Mary Falk. April 12, 2006. www.milkproduction.com (search article title).

FDA Product Recalls: The Role of FDA/The Responsibilities of the Food Industry, by Richard Silverman, Hogan & Hartson LLP. Presented at the 2005 American Cheese Society Conference, July 20–23, 2005, Louisville, KY.

Food Safety in Farmstead Cheesemaking. Dairy Practices Council Guideline No. 100. www.dairypc.org.

Chapter 13
Cultivating Farm Stays in California, by Janet Momsen and Jill Donaldson. *California Community Topics,* No. 7, April 2001. http://groups.ucanr.org/CCP/files/34360.pdf.

Marketing Meat Animals Directly to Consumers, by William R. Henning. Penn State University Cooperative Extension. www.bedford.extension.psu.edu (search article title).

Appendix C
American Raw Milk Cheese Presidium Protocols. www.rawmilkcheese.org/index_files/PresidiumProtocol.htm.

Bacteria Counts in Raw Milk, by Richard Wallace. University of Illinois Extension, Illini DairyNet Quality Milk Issues Papers, 2008. www.livestocktrail.uiuc.edu (search article title).

Milk Urea Nitrogen, by George Cudoc. Dairy One. www.dairyone.com 15, November 2009.

Raw Milk Quality Tests. Dairy Practices Council Guideline No. 21. www.dairypc.org.

Troubleshooting High Bacteria Counts of Raw Milk. Dairy Practices Council Guideline No. 24. www.dairypc.org.

Index

Note: page numbers followed by f, p, or t refer to figures, photographs, or tables

ADA (Americans with Disabilities Act), 158
advertising, 34
aging rooms/caves
    air quality, 131–32, 136–38
    building materials and systems, 142–50
    equipment, 133–38
    floor plan and utilities, 132–33
    humidity, 130–31
    photographs, 149
    shelving, 138–42
    temperature, 126, 128–30
Agri-Business Council of Oregon, 51
Agri-Services Agency, 51
agri-tainment, 174–79
agri-tourism, 173, 174–79
air quality, 131–32, 136–38
American Cheese Society, 5, 6, 7, 14, 51
American Farmstead Cheese (Kindstedt), 5
American Raw Milk Cheese Presidium, 5
Americans with Disabilities Act (ADA), 158
angel capital, 62
Animal Waste Management Plan (AWMP), 21
antechambers, 132–33
antibiotic residue test kits, 119–20
anti-siphon devices, 68–69
artisan cheese, defined, 6
ATTRA (National Sustainable Agriculture Information Service), 63
automobile insurance, 50
AWMP (Animal Waste Management Plan), 21

backflow prevention, 68–69
backflushing, 91
back-to-the-landers, 4
bathrooms, 157–58
B&Bs, 177–78
Bice, Jennifer Lynn, 4p, 5

biological oxidation demand (BOD), 74
Birchenough, Alyce, 14, 30, 84, 113p, 122, 162p
Black Mesa Ranch, 173, 174p, 190p
Black Sheep Creamery, 87p, 114p, 191
blogging/writing, 191–93
Blue Heron Farm, 182
BOD (biological oxidation demand), 74
boilers, 70p
Bonnie Blue Farm, 42, 137p, 178p
bookkeeping, 41, 53–56
bottling milk, 184–85
brand identity, 33
breaks, 189–90
brine tanks, 119, 133
brochures, 33
Bryant, Cary, 3
bucket milkers, 89–92, 100
budgets, 41–42, 180
building contractors, 20
building plans/permits, 18–20
buildings/structures, 79–81
bulk tanks, 97–100
business acumen, 11
business name, 34–36
business plans, 37–43
business structure, 40–41, 43–46
buying supplemental milk, 183, 213–14

CAFO (Concentrated Animal Feeding Operation), 20–22, 73, 76
Caldwell, Amelia, 57p
Caldwell, Gianaclis, xiip
Caldwell, Phoebe, 177p
Caldwell, Vern, 54p, 192p
California Mastitis Test (CMT), 209
Cal Poly Pomona, 14
candy, 185–86
cans and pails, 103
Capriole, 5
Carlson, Laurie, 75
Carlson, Terry, 26p

Carroll, Ricki, ix, xii, 5
carts, 120
CCPs (critical control points), 168–69
Cedar Grove Cheese, 74
ceilings. *See* walls and ceilings
certifications, 34
CFLs (compact florescent lights), 109, 130, 133
challenges, 12–13
Cheddar, 2, 74, 110t, 117, 119
Cheese and Fermented Milk Foods (Kosikowski), 5
cheese curds, 184
cheese guilds, 191
The Cheesemakers Manual (Morris), ix
cheese mites, 151
cheese presses, 117, 118p
Cheeses of the World (Nantet), 1
cheese vats, 111–14
chefs, 27, 30
chemicals, 102
Chenel, Laura, 5
Chez Panisse, 4
Child, Julia, 4
chilled water lines, 134–35
chlorine, 67, 167p
classes, 172–74
cleaned-in-place (CIP) systems, 90, 93, 100–01
cleaned-out-of-place (COP) wash sinks, 122
cleaning and maintenance, 91, 96, 101p, 102
Clouser, Gabe, 60
Clouser, Robin, 60
Cloutier, Pierre, 150p
CMT (California Mastitis Test), 209
Coleby, Pat, 189
coliform bacteria, 68, 210
Coliform Bacteria Count, 207, 208t
Columbus, Christopher, 1
combination vat and pasteurizer, 113–14
community-supported agriculture (CSA), 62
compact florescent lights (CFLs), 109, 130, 133
company name, 34–36
competitions, 27, 34
Concentrated Animal Feeding Operation (CAFO), 20–22, 73, 76

concrete culverts, 148
concrete masonry units (CMUs), 150
confections, 185–86
conferences, 191
Consider Bardwell Farm, 127p, 135p
construction and maintenance standards, 83–85, 94, 106–10
consumers, 25–28
contractors, 20
control measures, 168–69
CoolBot, 135
coolers, 133–35, 143–44
Cooperative Extension, 63
cooperatives, 3
copper vats, 113
COP (cleaned-out-of-place) wash sinks, 122
costs. *See also* financing; pricing
    certification, 34
    cheese press, 117
    cheese vats, 114
    energy/power, 76–79, 186
    environmental (EC), 80
    hidden, 186–87
    insurance, 50–51
    labor, 52–56, 186
    pasteurizers, 116
    product losses, 56–58
    start-up, 41–42
    steam kettles, 114
    total, 59–60
    triple-sinks, 121
creamery size, 7t
CreamLine, 69
creativity, 11–12
critical control points (CCPs), 168–69
CSA (community-supported agriculture), 62

Dairy Herd Improvement (DHI) milk testing, 187, 209
dairy inspectors, 17–18
Dairy Processing Methods to Reduce Water Use and Liquid Waste (Rausch and Powell), 69
days off, 190
DBAs, 44
DHI (Dairy Herd Improvement) milk testing, 187
direct sales, 28–29, 154–57

dishwashing, 12, 122
distributors, 30
DOE (U.S. Department of Energy), 72
donations to charity, 58
doors and openings, 85, 95
draining racks, 118
draining tables, 73p, 117–18
drains, 108, 121
drying racks/room, 118–19
Dunaway, Vicki, 69

EcoFlex Packager, 182, 183p
egg coolers, 134
Einstein, Albert, 188
electrical equipment, 108–09
electric power. *See* energy/power
Emery, Carla, 14
Employment Standards Administration (ESA), 55
The Encyclopedia of Country Living (Emery), 14
energy/power, 76–79, 186
environmental costs (EC), 80
Environmental Protection Agency (EPA), 20
ergonomics, 81
ESA (Employment Standards Administration), 55
Estrella, Anthony, 76
Estrella Family Creamery, 76, 149p
Estrella, Kelli, 76
estrus, 182
exhaust fans, 85, 95, 109, 136–38
exit strategy, 42–43

The Fabrication of Farmstead Goat Cheese (Le Jaouen), 126, 128
Fair Labor Standards Act, 54
Fairview Farm, 26p, 75p
Farm Credit System (FCS) Foundation, 63
farm dinners, 179
Farmers' Health Cooperative, 51
farmers' markets, 26p, 29, 179–81
farm insurance, 47, 49
Farm Services Agency (FSA), 60
farm stand/store, 28, 154–57
farm stays, 177–78
farmstead, defined, 6–7
fatigue, 12

FDA (Food and Drug Administration), 170
Federal Standards of Identity: The United States Code of Federal Regulations, 32, 164, 166
financing, 41–42, 60–64. *See also* costs
floor plans
　aging rooms, 132–33
　make room, 104–06
　Mama Terra Micro Creamery, 203
　milkhouse, 94
　milking parlor, 83
　Pholia Farm, 205
　small creamery, 204
　small non-licensed home dairy, 206
floors, 84, 94–95, 107. *See also* drains
florist coolers, 143
flow charts, 164–65
fluid milk, 184–85
flyers, 33
Fondiller, Laini, 105p, 180p
Food and Drug Administration (FDA), 170
foodies, 4
foot-pedal-operated sinks, 121–22
Foxfire books, 4
Fraga Farm, 55, 92p, 116
freezing milk products, 184
fromage blanc, xii
FSA (Farm Services Agency), 60
full-service cheese counters, 29–30
funding. *See* financing
Futhy, Brian, 189

good manufacturing procedures (GMPs), 92, 166–68
Gothberg Farm, 89p, 143p
Gothberg, Rhonda, 89p
Gouda, 2, 74
Grade A milk, 83, 107
Grade B milk, 85
grants, 62–63, 78–79
Grants.gov, 63
Gregory, Meg, 191
Gremmels, David, 3
grocery stores, 30

Hamby Dairy Supply, 83
Hamby, Paul, 83
hand-milking, 88

hand-washing sink and wash hose, 93–94, 97, 108, 121–22
Hazard Analysis and Critical Control Points (HACCP)
   critical control points, 168–69
   defined, 161
   hazard analysis, 168
   prerequisite programs, 164–68
   principles of, 162–63
   recalls, 169–71
headgates, 88
health insurance, 51
heating and cooling, 106, 126, 128–30, 133–35
Heininger, David, 173, 174p
Heininger, Kathryn, 173
HELOC (home equity line of credit), 60–61
herd size/productivity, 186–87
high temperature short time (HTST) pasteurizers, 116
hippies, 4
historical overview, 1–6
hobby cheesemaking, 8–9
holding tanks, 74–75
Home Cheese Making (Carroll), ix, xii
home equity line of credit (HELOC), 60–61
Hooper, Allison, 5
HTST (high temperature short time) pasteurizers, 116
humidifiers, 136
humidity, 130–31

income. *See also* costs; farmers' markets; financing
   agri-tourism, 173–79
   below minimum wage, 13
   classes/mentoring, 172–74
   increasing/prolonging production, 181–84
   meat sales, 185
   milk sales, 184–85
   outside sources of, 59
   sales, 28–30, 184–86
   value-added products, 9, 172, 185–87
Industrial Revolution, 2
insects, 150–52
insurance, 47–52
Internal Revenue Service, 53

interns, 52–56
Internships in Sustainable Farming: A Handbook for Farmers (Jones), 55
investors, 61–62

Jackson, Sally, 5
Jones, Doug, 55
Jones, Rodney, 39

Kill A Watt, 77
Kindstedt, Paul, 5
Kipe, Frank, 182
Kosikowski, Frank V., 5, 6

labeling, 32–33
Laboratory Pasteurization Count (LPC), 207, 208t
labor issues, 52–56, 186
laundry room, 157
Laura Chenel's Chèvre, 5
Lazy Lady Farm, 102p, 105p, 180p
learning, 14–15
LED lighting, 109
legal issues, 32, 52–56, 154f
Le Jaouen, Jean-Claude, 126, 128
liability insurance, 49
life insurance, 52
lighting, 85, 95, 109, 130, 133
limited liability companies (LLC), 44, 62
LLC (limited liability company), 44, 62
loans, 60–61
local market, 25–27
lotions and soaps, 185–86
LPC (Laboratory Pasteurization Count), 207, 208t

make room
   accessories, 118–24
   cheese press, 117
   cheese vat, 111
   combination vat and pasteurizer, 113–14
   construction and maintenance standards, 106–10
   draining table, 117–18
   floor plan, 104–6
   pasteurizer, 115–16
   steam kettles, 114–15
Mama Terra Micro Creamery, 60, 61p, 99p, 109p, 203

market research
   in business plan, 40
   people, 25–28
   place, 28–30
   pricing, 30–32
   product, 28
   promotion, 32–34
Mastering the Art of French Cooking (Child), 4
mastitis, 91, 209
McCalman, Max, 5
meat sales, 185
media coverage, 33
medications, 158
mentoring, 174
MicroDairy Designs, 182
milk cooling (bulk) tanks, 97–100
milkhouse
   construction and maintenance, 94
   equipment and accessories, 96–97
   floor plan, 94
   floors, 94–95
   hot water, 96
   lighting, 95
   milk cooling (bulk) tank, 97–100
   milk line washing equipment, 100–1
   strainer and receiving pails, 101, 103
   ventilation, 95
   walls, ceilings, and doors, 95
milking parlors
   construction and maintenance standards, 83–85
   equipment and accessories, 85
   floor plan, 83
   hand-washing sink and wash hose, 93–94
   milking systems, 88–93
   milk stands/platforms, 86–88
milk line washing equipment, 100–1
milk purchase agreement, 213–14
milk sales, 184–85
Milk Urea Nitrogen (MUN), 208
minimum wage, 13, 54
Minnesota Farmstead Cheese Project, 14
mission statements, 39
Monterey Jack, 2
Morris, Margaret, ix
Moss, Michael, 55
Mother Earth News, xii, 4
motivations for cheesemaking, 7–13

mozzarella, xii
MUN (Milk Urea Nitrogen), 208

NACMCF (National Advisory Committee on Microbiological Criteria for Foods), 162
name of business, 34–36
Nantet, Bernard, 1
National Advisory Committee on Microbiological Criteria for Foods (NACMCF), 162
National Renewable Energy Laboratoy, 79
National Sustainable Agriculture Information Service (ATTRA), 63
Natural Goat Care (Coleby), 189
Neilson, Jan, 55, 116
Neilson, Larry, 116
New England Cheesemaking Supply, 5
newsletters, 33
Nicolau Farms, 122, 141p
Nicolau, Walter, 122, 141p

off-grid power systems, 77, 79p
Office of Workers Compensation Program (OWCP), 55
office space, 153
open house events, 176–77
Orb Weaver Farm, 51, 149p, 189
Oregon State University, 13p
organizational structure, 40–41, 43–46
OWCP (Office of Workers Compensation Program), 55
owner-builders, 20

packaging, 28, 32, 155, 182
packaging room, 153–54
paper towels, 157
parlors. *See* milking parlors
partnerships, 44, 62
passive solar design, 81
Pasteurized Milk Ordinance (PMO), 83, 85, 107, 115, 116
pasteurizers, 115–16
Pastoral Artisan Cheese, 29p
pathogens, 210
pests, 151–52
pH, 68, 96
Phinney, David, 84
Pholia Farm
   bucket washer, 100p

cheese press, 118p
draining tables, 73p
floor plans, 205
marketing, 28
motivations for cheesemaking, 12
off-grid power system, 77, 79p
open house, 176p, 177p
photographs, xiii, 8, 11, 19
product losses, 56
shelving, 139p, 140p, 142
steam kettle, 115p
water heating, 70p
PIC (Preliminary Incubation Count), 207, 208t
Picket, James, 2
pipeline systems, 90–93
plumbing, 68–69, 108
PMO (Pasteurized Milk Ordinance), 83, 85, 107, 115, 116
Pollack, Marion, 51, 189
portable bucket systems, 90–93, 100
positive pressure, 106
poured/sprayed concrete, 148, 150
Powell, G. Morgan, 69
Powell, Maud, 55
Powell, Tom, 55
power bills. *See* energy/power
Prairie Fruits Farm, 179p
Preliminary Incubation Count (PIC), 207, 208t
presses, 117, 118p
pricing, 30–32
products
    buying supplemental milk, 183, 213–14
    donations, 58
    freezing, 184
    increasing/prolonging, 181–84
    liability, 49
    losses, 56–58
    profit margin and, 186
    selling fluid milk, 184–85
    value-added, 9, 172, 185–87
promotion
    advertising, 34
    brochures, newsletters, flyers, 33
    competitions, 34
    labeling, 32–33
    media coverage, 33
    packaging, 28, 32
    websites, 33
property insurance, 49–50
psychrometers, 131p
PVC pipe, 122

quality assurance. *See* Hazard Analysis and Critical Control Points

Rausch, Kent, 69
Raw Milk Cheesemakers' Association (RMCA), 5
recalls, 169–71
Redwood Hill Farm, 4p, 5
reefer trailers, 143
Reese, Bob, 5
refrigeration units, 134
refrigerators/freezers, 120
regulations, 32, 154f, 164
relative humidity, 130–31
renewable energy (RE), 78–79
restaurants, 27, 30
retailers, 27, 30, 31–32
risk assessment, 42–43
RMCA (Raw Milk Cheesemakers' Association), 5
rodents, 152
Rogue Creamery, 3
R-value, 129

safety. *See* Hazard Analysis and Critical Control Points
sales, 28–30, 184–86
Sally Jackson Cheese, 5
samples, 57p
sanitation, 12, 92, 102, 164–68. *See also* cleaning and maintenance; wastewater
SARE (Sustainable Agriculture Research and Education), 63
scales, 154f
SCC (Somatic Cell Count), 208, 209
Schad, Judy, 5
sea cargo containers, 146
seasonal milking, 189
Seger, Lisa, 182
septic systems, 74
septic tanks, 146–47
shelving, 96, 123–24, 138–42
shipping, 155
Simmie, Jacob, 177p

sinks, 93–94, 97, 108, 120–22
SIPs (structural insulated panels), 145
skipper flies, 152
Slow Food, 5
Small Business Administration, 37–38, 41, 45–46, 60
Smart, Willow, 84
soaps and lotions, 185–86
solar energy, 79, 80
sole proprietorships, 43–44
Somatic Cell Count (SCC), 208, 209
SPC (Standard Plate Count), 207, 208t
specialty cheese, defined, 7
Standard Plate Count (SPC), 207, 208t
Standard Sanitation Operating Procedures (SSOPs), 92, 164–68
steam kettles, 114–15
Stone Meadow Farm, 189
strainers/filters, 101, 103
straw bale construction, 147–48
structural insulated panels (SIPs), 80, 145
structures/buildings, 79–81
suitability for cheesemaking, 7–13
supermarkets, 3
Susman, Marjorie, 51, 189
Sustainable Agriculture Research and Education (SARE), 63
Sweet Home Farm
   ceiling materials, 84
   cheese press, 117p
   cheese vat, 113p
   commercial dishwashers, 122
   cows, 82p
   drying racks, 123p
   farm store, 30, 156p
   milking parlor, 87p
   pH testing, 162p
   pipeline milk hoses, 101p
   teachers, 14
   whey removal, 121p

tables, 122–23
Tanner, Gayle, 42
Tanner, Jim, 42
tasting room, 154–57
tastings, 175
teaching, 173
temperature control, 126, 128–30
terroir, viii
testing, 187, 207–11
toilets, 157–58
tourists, 27
tours, 175–76
towels, 157
trade shows, 27
training, 14–15
travel, 191
Twig Farm, 45p, 98p, 111p

ultraviolet (UV) water purification, 68
University of California at Davis, 174
University of Guelph, 14
USDA Rural Development, 63
U.S. Department of Energy (DOE), 72
U.S. Department of Labor, 55
used equipment, 114

vacation days, 190
value-added products, 9, 172, 185–87
vats, 111–14
Vella Cheese, 3
Vella, Ignazio, 3
Vella, Tom, 3
ventilation, 85, 95, 109, 136–38
venture capital, 62
Vermont Butter and Cheese, 5
Vermont Cheese Council, 5
Vermont Institute for Artisan Cheese (VIAC), 14
vision statements, 39

walk-in coolers, 143
walls and ceilings, 84, 95, 107–8, 133
Washington State University, 14
wastewater, 20–22, 72–76
water. *See also* wastewater
   calculating needs, 69
   heating, 70–72
   in the milkhouse, 96
   quality, 67–69
   reducing use, 69, 71
Waters, Alice, 4
websites, 33
Werlin, Laura, 5
Western Sustainable Agriculture Research & Education (SARE) Farm Internship Curriculum and Handbook (Powell and Moss), 55

whey, 74, 121p
whitewash, 150
wholesale sales, 29–30
Williams, Jesse, 2
Willow Hill Farm, 84
wind energy, 79
window air conditioners, 135
windows, 80, 85, 95, 106, 133, 153
wine cellar cooling units, 134

Wisconsin Specialty Cheese Institute, 7
Wolbert, Doug, 14, 30, 84
workload, 188–90
workshops, 172–74
writing/blogging, 191–93

year-round milking/birthing, 181–83

zoning, 16–17

About the Author

Gianaclis Caldwell, along with her husband, Vern, and their teenage daughter, Amelia, owns Pholia Farm, situated in the verdant Rogue Valley of southern Oregon, where they make aged cheese from the milk of their Nigerian Dwarf goats. The twenty-three-acre, off-the-grid farm and forest have been in Caldwell's family since the 1940s. Caldwell's critically acclaimed cheeses have been featured in books, articles, and top-ten lists. She is a former nurse and mixed-media artist.

About the foreword author

Jeff Roberts is co-founder and principal consultant to the Vermont Institute for Artisan Cheese at the University of Vermont and lives in Montpelier, Vermont.

Notes

Notes

Notes

Chelsea Green Publishing is committed to preserving ancient forests and natural resources. We elected to print this title on 10-percent recycled paper, processed chlorine-free. As a result, for this printing, we have saved:

17 Trees (40' tall and 6-8" diameter)
7,896 Gallons of Wastewater
5 million BTUs Total Energy
479 Pounds of Solid Waste
1,639 Pounds of Greenhouse Gases

Chelsea Green Publishing made this paper choice because we are a member of the Green Press Initiative, a nonprofit program dedicated to supporting authors, publishers, and suppliers in their efforts to reduce their use of fiber obtained from endangered forests. For more information, visit www.greenpressinitiative.org.

Environmental impact estimates were made using the Environmental Defense Paper Calculator. For more information visit: www.papercalculator.org.

the politics and practice of sustainable living
CHELSEA GREEN PUBLISHING

Chelsea Green Publishing sees books as tools for effecting cultural change and seeks to empower citizens to participate in reclaiming our global commons and become its impassioned stewards. If you enjoyed *The Farmstead Creamery Advisor*, please consider these other great books related to cheesemaking and farming-related business.

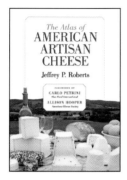

THE ATLAS OF
AMERICAN ARTISAN CHEESE
JEFFREY P. ROBERTS
9781933392349
Paperback • $35.00

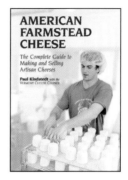

AMERICAN FARMSTEAD CHEESE
*The Complete Guide to Making
and Selling Artisan Cheeses*
PAUL KINDSTEDT
9781931498777
Hardcover • $40.00

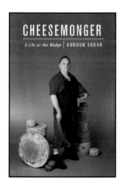

CHEESEMONGER
A Life on the Wedge
GORDON EDGAR
9781603582377
Paperback • $17.95

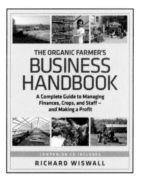

THE ORGANIC FARMER'S BUSINESS HANDBOOK
*A Complete Guide to Managing Finances,
Crops, and Staff—and Making a Profit*
RICHARD WISWALL
9781603581424
Paperback w/CD-ROM • $34.95

For more information or to request a catalog,
visit **www.chelseagreen.com** or
call toll-free **(800) 639-4099**.

the politics and practice of sustainable living

CHELSEA GREEN PUBLISHING

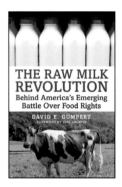

THE RAW MILK REVOLUTION
*Behind America's Emerging
Battle Over Food Rights*
DAVID E. GUMPERT
Foreword by JOEL SALATIN
ISBN 9781603582193
Paperback • $19.95

**INQUIRIES INTO THE
NATURE OF SLOW MONEY**
*Investing as if Food, Farms, and
Fertility Mattered*
WOODY TASCH
Foreword by CARLO PETRINI
ISBN 9781603582544
Paperback • $15.95

THE WINTER HARVEST HANDBOOK
*Year-Round Vegetable Production
Using Deep-Organic Techniques and
Unheated Greenhouses*
ELIOT COLEMAN
ISBN 9781603580816
Paperback • $29.95

SMALL-SCALE GRAIN RAISING, 2nd edition
*An Organic Guide to Growing,
Processing, and Using Nutritious Whole Grains
for Home Gardeners and Local Farmers*
GENE LOGSDON
ISBN 9781603580779
Paperback • $29.95

For more information or to request a catalog,
visit **www.chelseagreen.com** or
call toll-free **(800) 639-4099**.